Mule South to Tractor South

Two mules passing on a hill foreshadow the rise and fall of the mule in southern agriculture. (South Carolina Cooperative Extension Photographs, Special Collections, Clemson University Libraries)

Mule South to Tractor South

Mules, Machines, and the Transformation of the Cotton South

George B. Ellenberg

THE UNIVERSITY OF ALABAMA PRESS
Tuscaloosa

Hardcover edition published 2007.
Paperback edition published 2013.
eBook edition published 2009.

Typeface: Minion

∞

The paper on which this book is printed meets the minimum requirements of American National Standard for Information Science–Permanence of Paper for Printed Library Materials, ANSI Z39.48-1984.

Paperback ISBN: 978-0-8173-5772-6
eBook ISBN: 978-0-8173-8038-0

A previous edition of this book has been catalogued by the Library of Congress as follows:

Library of Congress Cataloging-in-Publication Data

Ellenberg, George B., 1958–
Mule South to tractor South : mules, machines, and the transformation of the cotton South / George B. Ellenberg.
p. cm.
Includes bibliographical references and index.
ISBN-13: 978-0-8173-1597-9 (cloth : alk. paper)
ISBN-10: 0-8173-1597-7 (alk. paper)
1. Mules—Southern States—History. 2. Farm tractors—Southern States—History. 3. Agricultural innovations—Southern States—History. I. Title.
SF362.E45 2008
631.3'710975—dc22

2007026071

To my mother and the memory of my father

Contents

Illustrations

Acknowledgments

Over the course of researching, writing, and revising this project, I have accumulated debts to many people and have enjoyed support in many ways. Much of the research for this study was funded by a University of Kentucky Dissertation Year Fellowship awarded by the Graduate School that helped provide a summer at the National Archives and Records Administration and the Library of Congress. In addition, many individuals have assisted in ways too numerous to list. The staff of the University of Alabama Press always responded to my questions promptly with clarity and good humor. They managed to get the project moving again and took pains to see it to completion. I appreciate the help that the readers provided to me which led to a stronger study. My copy editor, Lady Vowell Smith, provided extraordinary guidance in polishing the manuscript. I am in her debt.

Theda Perdue, under whose guidance this study was conceived, has always pushed me to do more than I believe I can. She never let the fire under me go out. I count it a privilege to have interviewed her father, Howard Perdue, who owned a Ford tractor dealership in Georgia. Thomas Cogswell, Ronald Eller, David Hamilton, and Dwight Billings offered helpful suggestions in the early stages of this project. Tom Appleton is to be acknowledged for his kind assistance in this endeavor, not to mention his continued interest in my work over the years. He has always been generous with his time and support. Jim Cobb, Pete Daniel, Gilbert Fite, Hardy Jackson, and Jack Temple Kirby encouraged me as a graduate student. I appreciate the scholarly examples they have set. Peter Coclanis has consistently shown a genuine interest in the project. Lu Ann Jones also labors in southern agricultural history and has been a source of encouragement over many years. Keith Harper and Bo Morgan have been unwavering in their support, patience, and friendship. They are the finest of colleagues and the best of friends.

Jane Halonen has provided a positive example in many ways and has taught me a great deal. She has also been unflagging in her support and has pushed me to take the time to focus on my scholarship. For that I am deeply grateful. Wes Little has been an example to me on many levels. I am glad to call him a colleague and a friend. Of special note are the support and encouragement that Charlie Mae Steen has supplied since my arrival at the University of West Florida, but especially in the past five years. She is the best in the world. Judy Jones has been instrumental in seeing to it that I had time to work on this project. Not only is she extremely capable, but her sense of humor always helps lighten the day. The staff in the Arts and Sciences Dean's Office makes going to work most pleasant. The members of the history department at the University of West Florida, especially Jay Clune and Dan Miller, always have been good colleagues, and Gabi Grosse has consistently been professional and supportive. The John C. Pace Library faculty and staff are all excellent in offering assistance, but I would like to thank Dan North in particular for the way he has shepherded the development of the southern history collection.

I would also like to thank the many librarians and archivists who were so helpful to me in my research. I am especially grateful to Dennis Taylor at Clemson University, the librarians and archivists of the Special Collections and Archives Department at the University of Kentucky, the archivists who oversee the Southern Historical Collection at the University of North Carolina, and the staff at the Mississippi Department of Archives and History. Thanks, as well, go to the Library of Congress staff, the archivists at the National Archives and Records Administration, the librarians at Mitchell Memorial Library at Mississippi State University, and Linda Coffman, registrar at American Mammoth Jackstock Registry.

Coleman J. and JoAnne Caudle Hardy are also gratefully acknowledged for their interest in this topic and their kindness in lending a family photograph for the cover illustration of this work. I would also like to extend thanks to Kenny and Renée Russell for their gentleness and hospitality in teaching people how to work with draft animals. Jimmy Klein is a master teacher in the draft-animal world, and I thank him for his good humor, patience, and thoughtfulness regarding my questions about farming with draft animals. It is people like them who are keeping draft-animal traditions alive and healthy. Although I have never met them, I want to thank Lynn Miller and those who publish *Small Farmer's Journal* for what they do for the draft-horse and mule community.

Over the years, I have always been able to depend on unending support from my mother, Claudia Latimer Bolton Ellenberg. She has been a rock every day of my life. She is one of the most intelligent and giving people I have ever known, and I count it a privilege to call her "Mama." I would also like to acknowledge my siblings, all five of whom have helped to shape my life and make it richer. Most especially, I thank my older sister Laura Ellenberg Gillespie for countless hours of help in my early graduate-student days and my older brother William Joseph Ellenberg for a place to stay whenever I needed it while researching at the Georgia State Archives during my master's program. To my sons, George Bolton Ellenberg Jr., Claude Easton Ellenberg, and Henry Fortson Ellenberg, I thank you for listening with such good humor and interest, for being yourselves, and for keeping my life in perspective. Finally, my wife, Karen Thomson Ellenberg, is the best friend and critic I have. Without her, this book would never have been completed. Words cannot express it, but she knows how much help she has been and continues to be every day.

Mule South to Tractor South

Introduction

Round and round the mule went, setting its narrow, deerlike feet delicately down in the hissing cane-pith, its neck bobbing limber as a section of rubber hose in the collar, with its trace-galled flanks and flopping, lifeless ears and its half-closed eyes drowsing venomously behind pale lids, apparently asleep with the monotony of its own motion. Some Homer of the cotton fields should sing the saga of the mule and of his place in the South. He it was, more than any other one creature or thing, who, steadfast to the land when all else faltered before the hopeless juggernaut of circumstance, impervious to conditions that broke men's hearts because of his venomous and patient preoccupation with the immediate present, won the prone South from beneath the iron heel of Reconstruction and taught it pride again through humility, and courage through adversity overcome; who accomplished the well-nigh impossible despite hopeless odds, by sheer and vindictive patience. Father and mother he does not resemble, sons and daughters he will never have; vindictive and patient (it is a known fact that he will labor ten years willingly and patiently for you, for the privilege of kicking you once); solitary but without pride, self-sufficient but without vanity; his voice is his own derision. Outcast and pariah, he has neither friend, wife, mistress, nor sweetheart; celibate, he is unscarred, possesses neither pillar nor desert cave, he is not assaulted by temptations nor flagellated by dreams nor assuaged by vision; faith, hope and charity are not his. Misanthropic, he labors six days without reward for one creature whom he hates, bound with chains to another whom he despises, and spends the seventh day kicking or being kicked by his fellows. Misunderstood even by that creature, the nigger who drives him, whose impulses and mental processes most closely resemble his, he performs alien actions in alien surroundings; he finds bread not only for a race, but for an entire form of behavior; meek, his inheritance is cooked away from him along with his soul in a glue factory. Ugly, untiring and perverse, he can be moved neither by reason, flattery, nor promise of reward; he performs his humble monotonous duties without complaint, and his meed is blows. Alive, he is haled through the world, an object of general derision; unwept, unhonored and unsung, he bleaches his awkward accusing bones among rusting cans and broken crockery and worn-out automobile tires on lonely hillsides while his flesh soars unawares against the blue in the craws of buzzards.

—William Faulkner, *Sartoris*

The mule has played a central role in southern life, mind, and letters. In developing metaphors for the South, for example, William Faulkner often employed the mule. In reality, Faulkner also dabbled in mule raising.[1] Jerry Leath Mills argues that a truly southern tale always contains a dead mule. His research led him "to conclude, without fear of refutation, that there is indeed a single, simple, litmus-like test for the quality of southernness in literature, one easily formulated into a question to be asked of any literary text and whose answer may be taken as definitive, delimiting, and final. The test is: *Is there a dead mule in it?*" As Mills points out, "the presence of one or more specimens of *Equus caballus x asinus (defunctus)* constitutes the truly catalytic element, the straw that stirs the strong and heady julep of literary tradition in the American South."[2] For the historian, as well as the novelist, the mule offers an intriguing means of exploring the regional identity of the South, its singular heritage, and the profound changes it has experienced.[3] A special relationship exists between the American South and the mule. This study traces the journey of the mule through southern culture and agriculture. Colorful characters such as politicians, mule traders, farmers, editors, and bureaucrats enliven the story, but the mule is the most interesting character of all.

Ironically, the animal's very ubiquity camouflaged its importance to the South. It was so much a part of the fabric, sights, smells, and sounds of southern life, so much a given, that its presence dulled observers to its significance—because it was so *southern*. This is true in the broader sense of draft animals in American history as well, partly because technologies made of iron and steel have overshadowed the central importance of draft animals to both the agricultural and industrial sectors well into the twentieth century. Studies of horses and mules are needed to understand fully both particular and broader historical issues related, for example, to technology and society.[4] "Horses lack both history and historicity," argues one scholar, and the same holds true for mules.[5] By using mules as a touchstone, we can discover much about southern "social organization, values, behaviors, racial attitudes, and racial and class struggles" from the 1850s until the 1950s.[6] Mule stories abound, but that is not enough. Mules are an integral component of southern history, and the fact that southerners used mules instead of other draft animals tells us something about the region. The "shabby neglect of this admirable creature" has left an incomplete story of the region, and a much less interesting one as well.[7]

To understand the South, one must confront southern agriculture and its central place in the region's history. To understand southern agriculture, one must confront the mule, and this confrontation is the fundamental purpose of this study. Steven Stoll, in another context, put it this way—"farming matters."[8]

Fig. 1. Mules were ubiquitous in southern fields and southern towns, as this scene in Port Gibson, Mississippi, shows. (Courtesy of Mississippi Department of Archives and History)

The mule may be the best symbol of the agricultural South in many ways, although some may argue that point.[9] This, then, is the story of the South from a particular angle. If the new social history provided impetus for a view from the bottom up, then this perspective is from the back end of a cotton mule, a view that changed little from furrow to furrow, year to year. James C. Cobb notes that "[t]he history of southern identity is not a story of continuity *versus* change, but continuity *within* it."[10] This is an apt description of the mule's place in southern agriculture and culture; the mule was a constant for a century as the region changed around it. In the end, however, the region was transformed from the mule South into the tractor South.

This study has several major points. The discussion addresses the cotton-growing regions of the South. Mules were used for many purposes, but cotton was certainly one of the crops most closely tied to mules. Also, mules did not come to prominence in the South until the antebellum period, despite the hopes of men like George Washington. Mules were known to many southerners during the Revolutionary era, but widespread adoption did not take place until later. Mules were an example of how southern planters attempted to innovate within the confines of the slave regime. That is, mules provided a means of safely linking the South to certain elements of a progressive nation. In the strictest sense, however, mules were innovative but not necessarily progressive. It is important to note that it was planters, and not the yeomanry, who first adopted mules, although mule use became much more common after the end of the Civil War. It may be that mule ownership is one indicator of a farmer's relationship to cotton and slavery, as well as a possible indicator of his aspirations. The mule trade and mule raising are important to explore because the cotton South raised few mules. One aspect of the mule's place in history that is not addressed at length is that of mule use during the Civil War or the World Wars. Clearly, this is an important aspect of mule history, and all three wars have an impact on draft-animal populations, but it is a topic in and of itself which has been treated elsewhere.[11]

I also hope to illuminate some of the contradictions and debates that occurred as the nation's farmers grappled with the enormous changes catalyzed by the machine age during the early and mid-twentieth century. In the context of mechanization, there is a point at which the study leaves the South in order to examine the national background of farm mechanization, which in turn had a tremendous impact on southern agriculture. As tractors came to dominate the Corn Belt, much of what happened there influenced the South in terms of the evolution of United States Department of Agriculture (USDA) attitudes and policies, as well as the southern mule supply. The story of the Corn Belt illustrates how closely the cotton South was linked to areas outside its borders for its mule supply.

In the wake of midwestern tractorization, southern planters looked hard at the shortcomings and possibilities contained in fully mechanized cotton growing, and that exercise was tied intimately to social relations in the region. While the Department of Agriculture played an important role in mechanizing the American farm, agricultural magazines, implement manufacturers, and tractor dealers also contributed significantly to the changes that rippled outward. Public and private institutions were important components of the fundamental shifts that occurred in the mid-twentieth-century South. Finally, and perhaps most importantly, I hope this work will offer insight into the place of the mule in southern life and culture and, to a small extent, link it to some broader agrarian themes. "Mules are," as Harry Watson notes, "emblematic of the South's rural past."[12]

This study does not spring from a nostalgic desire to return to the past, and should not be read as such. It is an exploration of the past. The two are not the same, although they are often conflated when one begins writing in terms of choices that could have been made. As Wendell Berry reminds us in *The Gift of Good Land,* these choices rested on "our willingness to limit our desires as well as the scale and kind of technology we use to satisfy them. Without that willingness, there is no choice; we must simply abandon ourselves to what the technologists may discover to be possible."[13] In an age in which there is a "gulf of incomprehension between shopping cart and pasture" and parents explain to children that the cows they see at fairs are not the cows that they eat, it may be healthy to look a little more carefully at the world that used to exist.[14] The same holds true of the region's history; many people do not know how different the South used to be. In the case of the South, there are things to be proudly held up for praise, and there are things that we wish could be hidden. Both are integral components of a past in which mules were central.

Once, almost every southerner, black or white, male or female, young or old, knew something about mules, and probably a good deal. That is no longer true. The mule is a hybrid animal, the offspring of a male ass, or jack, and a female horse, or mare. Both male and female mules result from this union. The reverse of this cross, a stallion coupled with a female ass, produces a hinny, which in the United States was considered inferior to the mule as a work animal. The mule, because of a genetic phenomenon known as hybrid vigor, is often larger than either of his parents.[15] Allegedly, the hinny does not receive the genetic vigor that a mule does, although modern theories of heredity counter the commonly held view of breeders that hinnies are inferior in quality and size.[16] Few hinnies were produced. Mules do possess the drive to mate, so male mules were castrated as a matter of course because the operation made them more tractable. Even so, female or mare mules were generally preferred for field work. Because particulate matter trapped in a male mule's sexual organ could cause discomfort and irritation, it would make him hard to handle.[17]

Mare mules very occasionally have been reported to conceive and carry a foal to term, but there is no scientific support for such claims.[18] The main point is this—both asses and horses must be used to produce mules. Asses have distinctive traits which are passed to mules. Perhaps the most obvious are the large, long ears which are the signature of mules. Asses have much less mane and tail hair. Mule tails have little hair along the first several inches. Mules also inherit the ass's thick back and rib skin, making them less prone to suffer from ill use, punishment, and biting insects.[19] Asses also contribute steadiness, surefootedness, and hardiness under adversity; horses contribute size and alertness.[20]

The American Jack, ultimately a breed with its own registry, resulted from the mixture of several jackass breeds, all with southern European and Mediterranean roots.[21] European breeders made efforts to keep the various strains separate, but more practical and less tradition-bound American breeders crossed and blended the various bloodlines, sometimes with the thought of developing a superior breeding stock exhibiting the best attributes of each European breed and sometimes as a result of careless breeding. The result, the American Jack, is both distinct from its several parts and more useful for the raising of mules than any other jack breed.

American breeders often crossed various strains of jacks to bring out desired characteristics, but the most significant breed imported into the United States was the Catalonian, with the Andalusian standing as next important. Most Catalonians possessed a glossy, black coat, although other colors were not unknown. Standing 14½ to 15 hands as a rule (a hand being four inches with height of animal being measured at the withers), a few reached 16 hands in height, and they purportedly had great style, beauty, and superb action. Breeders treasured the Catalonian's liveliness as a curative for the nearly universal lack of fire in American jacks and jennets who acted as though playing in a pasture was "far beneath their sense of dignity and decorum." Early breeders believed color to be a key indicator of a mule's worth, and Catalonians produced more dark colts than did many other breeds and were therefore highly prized.[22] Henry Clay experimented with the Catalonian breed and imported a number of the animals from their native region in northern Spain.

Another of the most sought-after European breeds was the Andalusian. George Washington and Henry Clay lent their respective statures to the Andalusian breed, thus encouraging much experimentation with it in the United States. Andalusians, natives of southern Spain, dated back at least as far as ancient Rome. They measured between 14½ and 15 hands and most often were gray, although nearly white examples were noted, and black and even blue ones were not unknown. Their legs were large and firm, and their heads and ears were considered moderately well shaped.

Maltese jacks came into America around the same time as Andalusians, in time becoming perhaps more popular than the latter. They bear the name of their home in the Mediterranean and were small, hardly ever standing over 14½ hands, with the average Maltese jack measuring around 14 hands. Unlike Andalusians, Maltese jacks were black or brown. Brown was the signature color of pure-blooded animals. They were known for their well-shaped heads and upright ears, as well as for their vitality. Often, however, their legs were small boned, thus weakening their appeal for breeders pursuing size and power in their animals.

Larger than the Catalonian, but often quite as popular, the Majorcan breed was valued primarily because of its size. Majorcans stood on average a hand taller than Catalonians. Because of the limited supply of this breed, they did not have as great an impact as some of the other breeds. Their heads and ears often appeared bulky, and they had the reputation of being sluggish. They normally had black coats, although they were not as glossy as the Catalonian breed.

Two other breeds deserve mention, the first less important than the second. Italian jacks constituted a geographical breed rather than a genetically unique breed. Of mixed ancestry and small, the Italian breed also cost less than other breeds. Because they were so small, usually between thirteen and fourteen hands, American breeders considered them unfit for siring jacks. The Poitou ranked as high as the Italian ranked low. A French breed, the Poitou's most distinguishing characteristic was its thick, long coat. Seldom groomed in its native land, the Poitou did not at first glance give a very favorable impression to American sensibilities. Even so, Europeans believed "the mules from Poitou are the largest, heaviest and best to be seen in Europe." European demand for the Poitou was great and limited the impact the breed had on the American mule, although some Poitou blood may have added to the size of the American Mammoth breed.[23]

In the American context, mules fell into five general market classes—mining mules, cotton mules, sugar mules, farm mules, and draft mules. Mule size varied widely, usually ranging from six hundred to eighteen hundred pounds. Mining mules, obviously purchased for use in and around mines, were compact. Small mining mules were called pit mules. Cotton mules generally weighed eight hundred to one thousand pounds, although many were probably on the lighter end of the scale. Most worked on cotton farms and plantations, but they were also widely employed to pull delivery wagons in many cities. Sugar mules worked on sugar plantations in the Deep South and were larger than cotton or mining mules. Sugar mules might stand sixteen to seventeen hands high and weigh 1,150 to 1,300 pounds. Farm mules possessed neither uniform conformation nor style and finish. They were used for general farmwork. Draft mules were the largest mules and ranged in height from 16 to 17½ hands and weighed between 1,200 and 1,600 pounds, with a few animals reaching a slightly larger weight.[24] The larger animals worked in the heavy soils of the South, as well as in southern forests, where they hauled trees. The smaller animals worked in areas more suited to their size and power, such as in areas of lighter soils, and smaller animals meant less required feed, smaller harness, and generally lower upkeep costs.

Long before mules became popular in the United States, they had attained noble stature in other parts of the world. In the ancient world, mules carried kings, generals, standard-bearers, trumpeters, and drummers into battle. Because mules often stood taller than ancient horses, they had obvious advantages in battle. Commanders could see more of the battlefield, and troops could better hear and see those persons who carried colors and orders. The Greeks prized mules and employed them in civilian life to pull carts but reserved horses for their war chariots.[25] Mules pulled Alexander the Great's funeral chariot.[26] The Romans, too, used mules to pull carriages and as pack animals in military-supply trains. Although the Israelites did not breed mules, they used them widely, purchasing them from suppliers such as the Hittites.

Mules in northern Europe never achieved the same status as those south and east. Eschewed by European nobility, mules plodded along as pack or plow animals, or as mounts for priests and ladies of rank. The clergy's link to mules may stem from the ancient Jewish predilection for mules and asses.[27] Broadly speaking, southern Europeans of the Iberian peninsula, southern France, Italy, and to a lesser extent the Balkans employed mules much more than did other Europeans.[28]

Although mules were widely used in the American South, the animals never achieved high status. In the South, mules remained hitched, not to the chariot but to the plow and wagon, all the while subjected to broiling summer heat, poor care and feeding, and the lash. Still, if it is "interesting to speculate how different the course of Western civilization might have been without the invention of the mule," it is equally interesting to imagine how the American South could have developed as it did without the mule.[29] The animal that William Faulkner called "anonymous avatar of intractable Mule" patiently bore his lot in southern life for more than a century, emerging in the process as a foundation and symbol of the region's agriculture and culture.[30]

1
The Mule Ascendant
Oxen, Horses, and Mules

Commenting on the history of jacks and mules in the American South, the compilers of the *Twelfth Census of the United States* made the telling observation that during the 1830s, "the prejudice against mules finally gave way."[1] Indeed it did, for over the next three decades mule use expanded dramatically in the South, and by the eve of the Civil War mules had joined land, slaves, and cotton as a standard fixture of southern agriculture.[2] The South's adoption of the mule did not come easily or quickly, as the debate that peppered the pages of antebellum agricultural journals illustrates. But the mule's steady progress in gaining acceptability signaled the willingness of many southern planters to embrace scientific farming in the form of selective breeding. In today's terms, one could say that southerners adopted a draft animal produced by a "particularly specialized form of biotechnological practice."[3] Planters may have wondered at the terminology, but they would have understood the mule's distinctiveness.

The adoption of the mule, along with the disinclination of southerners to produce them locally, demonstrated that many of the region's inhabitants stood willing to participate in a national and even international economy beyond the production of staple crops. The mule became increasingly and to a large extent uniquely southern, but at the same time, it linked the South to the nation ideologically. The mule, however, also revealed divisions within southern society, and while antebellum southerners popularly associated mules with slaves, a more appropriate association might have linked mules with status, wealth, and power, since it was the plantation owners who first experimented with and then widely adopted mules to be used by their slaves.

During the early and mid-nineteenth century, the nation as a whole underwent significant changes. The transportation and communication revolutions began to reshape the landscape and the way people viewed the world. Politics exhibited its own paroxysms as the Revolutionary guard passed away and a new generation wrestled with societal changes and political pressures. For all the upheavals, however, the United States remained largely a land of agriculture and farmers.

Innovations and technological changes reverberated through agriculture, just as they did through other sectors of society. Whitney's cotton gin provided much of the impetus for an unprecedented explosion of cotton cultivation in the upland South which subsequently swept across much of the region. As Wilbur Cash observes in *The Mind of the South,* "Cotton would release the plantation from the narrow confines of the coastlands and the tobacco belt, and stamp it as the reigning pattern on all the country. Cotton would end stagnation, beat back the wilderness, mow the forest, pour black men and plows and mules along the Yazoo and the Arkansas, spin out the railroad, freight the yellow waters of the Mississippi with panting stern wheelers—in brief, create the great South."[4] Yet Whitney's invention applied directly to processing the cotton fiber, not cultivating and harvesting the crop. What went on in southern cotton fields remained mostly untouched by technological innovation, largely due to the need to keep laborers occupied throughout as much of the year as possible.

In the antebellum South, agricultural reform presented its proponents and converts with potentially unsettling implications and possibly revolutionary repercussions. Planters could not tolerate reform that might threaten a status quo that rested on slavery. Thus, the paradox of southern agricultural reform centered upon, as did much Old South thinking, its impact on the institution of slavery. Southern planters found themselves bound to an ancient labor system which by 1830 had been granted a sacrosanct position within the southern value system and which, in many ways, limited agricultural reform.[5]

As tensions heightened between the sections after 1830, the South remained optimistic about the viability of slavery and the future of cotton production. Editors of and correspondents to the several southern agricultural journals revealed few misgivings about the future, and remained sanguine, especially about the potential for cotton cultivation, up to the Civil War. Looking back on this period, William J. Cooper asserts, "One word characterizes the agricultural and particularly the cotton outlook in all the journals: optimism!"[6] At the same time, southerners revealed a growing sensitivity to outside attacks, especially those that questioned the vitality of southern civilization or the morality of southern institutions.

Wrapped within the debate about agricultural reform, to which the mule belonged, lay the issue of the place and role of the South in the United States, and the ability of the section to compete in a "modern" world. Between 1820 and 1840, for example, railroad entrepreneurs laid twenty-six thousand miles of track, while Samuel F. B. Morse made American telegraphy a reality in the 1830s.[7] The textile industry, to which the South was intimately joined, stood as the "most fully mechanized industry" in the world.[8] Southern planters, like mules, were hybrids of a sort, harnessed to an ages-old labor system, but also closely tied to a developing economy and themselves flirting with agricultural innovation.

Southerners did not consider slavery and progress as mutually exclusive and attempted to embrace elements of the modern world without compromising the foundations of the South's distinctiveness and, in its own eyes, its superiority.[9] As Mark Smith notes, "while southern masters courted some of the most economically capitalist features of the nineteenth-century world, they refused to endorse those that threatened the integrity of their society."[10] By the same token, they embraced those that offered to strengthen their society. While planters were not totally convinced that the modern world held all the answers to their problems, they did believe that they could properly balance the demands of controlling slavery and looking forward.[11] It was, of course, a tightrope fraught with difficulty since missteps could have incalculable repercussions.

The growing interest in mules sprang from a larger concern with agricultural innovations such as improved plows, the use of fertilizer, and a desire for greater efficiency on the part of farmers. "Time is everything," noted Hill Carter of Virginia.[12] And to some, time went hand in hand with a shift to mules. "Some of our practical and sagacious farmers have commenced with the work of reform," the *Southern Planter* commented in 1852, and "have dispensed with the horse and supplied themselves with mules."[13]

At the least, mules represented innovation in the sense that they were the least traditional of the three types of draft animals, but shifting to mules did not necessarily make a farmer innovative in every respect. Mules could be, and were, used with the most backward farming tools and techniques, as well as with the more advanced, as the state of southern agriculture in the twentieth century showed. Yet, during the antebellum years, mules represented a departure of sorts from well-worn customs, and planters who owned mules often hitched them to improved agricultural implements. Conversely, farmers who employed oxen owned few improved tillage tools.[14]

Many experiments during the 1840s and 1850s increased the information available to interested farmers about the draft requirements of various implements. Farmers often remained tied to time-tested implements or techniques which might be wasteful, simply because they knew they could at least produce a crop with familiar methods. One writer complained that farmers as a lot "do not actually understand the principles of a plough, are not aware when it is out of order, and could not for the soul of them superintend the grinding of an axe."[15] An editor noted that farmers who "generally, are little aware of the very great differences in ploughs . . . have provided themselves with patent churns, and patent beehives of the latest construction, [but] they continue to use a plough, the idea of which may have been obtained from the Georgics of Virgil, or from the sample preserved in the Chinese exhibition."[16]

The mule offered the best of all possible options for planters. The mule first found acceptance among planters who on the one hand boasted of the superiority of their labor and agricultural systems, but on the other hand yearned to demonstrate that they could innovate within the context of the plantation regime. In a way, planters needed to prove they could exist and "compete" in the modern world. As Mark Smith states, "planters wanted to be perceived as modern, and they wanted to make money."[17] As the members of southern society at once most closely shackled to slavery as their source of livelihood and to the market economy, and most alert to criticisms of the region and its agricultural practices, planters also possessed the capital with which to purchase and experiment with the more expensive mule.

Scattered numbers of mules could be found in the South before Whitney patented the cotton gin, but not until later did the animals become numerous. In North Carolina mules first appeared in the 1790s or early 1800s, and their numbers increased by more than one hundred percent between 1850 and 1860. Despite the rapid advance in mule numbers, by the latter date North Carolinians owned three times more horses than mules or oxen.[18] Mississippi planters began shifting more to mules after 1840, although oxen and horses continued to be used throughout the antebellum period.[19] In Louisiana's sugar country, oxen, which had been the draft animal of choice for planters during the early nineteenth century, were replaced by horses and mules. By the late antebellum era mules had become the primary draft animals on sugar plantations.[20] By 1860 in Georgia, mules outnumbered horses in 31 of 132 counties.[21] At the same time, oxen outnumbered mules in sixty-one counties, and oxen outnumbered horses and mules in three counties.[22]

The number of mules in the South more than doubled between 1850 and 1860, rising from 405,222 to 822,047. Despite this large increase in mules, however, southern farmers still owned more oxen than mules on the eve of the Civil War, but the percentage rate of growth for oxen between 1850 and 1860 was less than half that of mules. In Georgia, for example, large plantation owners tended to replace oxen with mules after 1840, but the total number of oxen

Fig. 2. Flat fields such as this one in Mississippi were well suited to large numbers of mules and, later, to mechanization. (Courtesy of Mississippi Department of Archives and History)

in the state still increased between 1850 and 1860.[23] Horses outnumbered mules and oxen in the South throughout the antebellum period, but their relative popularity declined between 1850 and 1860. The number of oxen rose from 603,594 to 858,494 during the decade, while the draft-horse population rose from 1,421,014 in 1850 to 1,763,697 in 1860.[24]

In 1850, mules accounted for seventeen percent of southern draft animals. Horses made up fifty-eight percent of the total and oxen the remaining twenty-five percent. Mules accounted for only twenty-four percent of southern draft animals in 1860, despite rapid increases in their numbers. Horses accounted for fifty-one percent of the total and oxen for the remaining twenty-five percent.[25] Despite its numerical dominance, however, the horse was the only draft animal whose percentage declined during the decade before the Civil War.

The debate over draft animals occurred during a period in which larger agricultural issues related to land use became important. From the 1820s to the 1840s, the question was: "What combination of crops and animals would most likely result in an auspicious fusion, feed the larder and the account book, make the agriculture of the United States the 'great sources of its riches, and the right arm of its power,' and stave off a hunger for fresh starts?"[26] Southerners may have viewed the question and answer differently from their northern counterparts, but the question arose nevertheless in North and South.

So, an environment for agricultural improvement existed, but not all southern farmers and planters readily or hurriedly embraced mules. In late-eighteenth-century low-country South Carolina, more planters kept bees than owned mules; 1.3 percent of estates listed mules, while 7.6 percent owned bees.[27] Southerners examined their draft-animal options closely, since agricultural production rested on draft animals in the most fundamental way. Skeptics abounded, questioning the advisability of choosing the "new" draft animal over the more traditional, tested, and proven horse and ox, and a debate broke out in the pages of agricultural journals that was never fully concluded, although mules became much more common by the time of the Civil War.

It is impossible to identify the opening salvo in the debate, but it began around 1819 when a correspondent to the *American Farmer* wrote, "We have in our country only three species of beasts of burthen in common use, the horse, the ox, and the mule."[28] Horses and oxen did not disappear from southern farms during the antebellum years. Nor was it clear that mules would predominate throughout the southern agricultural sector. But by the eve of the Civil War, mules had a large and growing number of champions throughout the South, the infrastructure for raising and supplying the region with the animals had matured, and it was clear that mules would play an important role in the region's agriculture. By the closing years of the antebellum era, mules had achieved a prominent position in the cotton regime's draft-animal hierarchy.

Planters and farmers discussed at length their feelings and findings about draft animals in agricultural journals such as the *Southern Cultivator*, the *American Farmer*, and the *American Agriculturist*. The *Cultivator* was the premier farm journal of the antebellum South. Founded in 1843 in Augusta, Georgia, this journal advocated agricultural reform, especially diversification, improved livestock, and cover crops. The *American Farmer*, founded in 1819, was a Baltimore, Maryland, journal which focused on the eastern seaboard but also strove to be broader in geographical scope. The *American Agriculturist*, a New York periodical, added a northern voice to the discussion.[29] Even the influential *De Bow's Review* occasionally addressed the draft-animal issue in spite of the fact that agriculture was not its primary focus.

To be sure, not all southern planters and farmers read any or all of the journals, but the papers did have readers in the South. Planters and farmers often exchanged journals just as they traded gossip, news, and other information on the porch, in the square, and in front of the church. At the very least, neighbors would be interested in what was going on in one another's fields. The dynamics of southern class relationships aside, the views represented in the agricultural journals accurately reflect planter thought.[30]

The editors of these journals self-consciously and vigorously advocated agricultural reform, especially between 1820 and 1850.[31] The journals reported new plow designs, plow tests, proper hitching methods, machinery, and progressive farming techniques such as drainage, new cropping routines, manuring, farming different soil types, and other agricultural issues. Both northern and southern antebellum agricultural journals regularly carried discussions of these topics even if they also decried, from their perspective, the glacial pace of changes. It is not surprising that agricultural reform or "improvement" found a constituency beginning in the 1820s for many reasons, but among the most important was the fact that agriculture no longer appeared to be the foundation of the nation's economic, political, and social fabric.[32]

The *Southern Cultivator* recommended careful husbandry in place of inefficient practices, always stressing the point that to "work poor mules, oxen and horses, or waste their expensive food, is bad economy."[33] The journal also applauded the "spirit of inquiry" reflected in questions regarding sound farming practices.[34] The journals provided a sounding board for opinions and reports on particular experiences with draft animals. Clearly not scientifically accurate, especially in terms of animal diet and disease, the letters and reports nevertheless illuminate what readers thought about the different types of draft animals. In regard to the southern draft animal debate, then, perceptions and impressions often meant as much as "scientific" proof to those involved. At the very least, the dialogues demonstrate that southern planters were intensely interested in the draft-animal question.

Edward Stabler of Maryland, for example, substituted oxen for horses on his farm in 1822 or 1823. In 1850, he still advocated oxen, stating that for "many years, there was not a furrow plowed on his farm except by oxen."[35] Other farmers vociferously defended the mule. "I am well aware of the prejudice that exists against this humble, faithful and most abused servant of man; but I know him well and there is not a charge against which, I am not prepared to defend him," stated one mule supporter.[36] Horse advocates, too, penned similar lines. "The horse," opined one farmer, "is undoubtedly the most useful and manageable of all animals known to man."[37]

It is hardly surprising that farmers held horses in high esteem. In concert with tradition, the horse possessed class and speed. Horse advocates described their choice of animals as stylish, spirited, handsome, and adaptable. The horse, stated one writer in the *Southern Planter*, "works patiently and steadily at the plow, or in drawing the loaded carriage."[38] Another commentator attributed the horse's speed to superior physiology. In a burst of unfounded generalizations, he stated that "mules cannot compete with horses in point of speed. The material of which a mule is made seems to be tougher, and less given to motion, so that the effort a horse uses in making four miles an hour, a mule makes not more than three."[39]

Speed, style, and status went hand in hand. Even when a farmer employed oxen or mules for field work, he did not use them in his personal livery. Mules might have their own style, but that style was considered uniquely suited to the plow, not to the carriage. Even observers who marveled over the sight of a mule in the field, "gliding between the crop, touching on neither side, and walking about four miles an hour," found it inconceivable or improbable that mules could serve in a gentleman's hitch.[40] As the *Southern Cultivator* explained, the "wife of almost every planter who is doing well, keeps her carriage and a pair of horses . . . and every son and daughter big enough to ride, wants a saddle-horse besides."[41] Mules were fine for slaves and field work, but not for a planter's carriage, his spouse, or his offspring.

Status, however, had its price. Horses had the reputation of being costly and extravagant. To some, the desire to use horses added up to simple wastefulness. During the less than flush days of the early 1840s, this "natural and costly predilection" to excess, as the *Cultivator* proclaimed, might have been fine in a previous era, but it was "no more congenial with the utilitarian, money saving age of ours, than would be the vagaries of the Knight of LaMancha himself."[42] Nevertheless, many southerners were reluctant to part with their horses, tacitly echoing the sentiments of the individual who called the idea of using mules to draw a carriage "preposterous."[43]

Oxen were long familiar to farmers and retained a substantial following throughout the antebellum years, although one historian attributed the increase in mules in central Georgia to the possibility that slaves "were poor ox drivers."[44] In 1852, the Southern Central Agricultural Society of Georgia still offered a premium for the best yoke of oxen at its annual competition, and agricultural journals regularly discussed the beast within their pages.[45] In determining how best to compare oxen with other draft animals, the *Southern Planter* admitted the task to be "a very complicated and difficult one."[46] Oxen had many qualities that made them a sound choice for farmers. Above all else, the ox connoted dependability. A "patient, willing laborer," the ox was said to be "the most docile and tractable of three several species of beasts of burthen. If he has not the gaiety of the horse, neither has he that fretful restiveness of temper which renders the horse, at times, entirely unmanageable. Neither is the ox ever influenced by the sulky perverseness of the mule."[47]

Aside from allusions to the ox's character, advocates asserted that this beast could outwork horses and mules despite its faults, foremost among which was slowness. An almost universal complaint about the ox hinged on speed. While antebellum observers did not understand the physiological explanation for this, they were correct since, as ruminants, oxen extracted more energy from food, but at the cost of speed.[48] Industrious, patient, and sagacious the animal might be, but even in the antebellum South, time was important when preparing fields. Some admitted that the ox was slow, but argued that the real issue lay not in which animal was the fastest but which could perform the most "heavy labor in the shortest time and at the least expense."[49] To the consternation of some observers, oxen at times won the speed contest. At a timed plowing match in New Jersey, for example, it was reported that a yoke of oxen defeated a team of horses "again." The results of the contest confounded the *Southern Planter*, which concluded that "either our horses at the South are much better, or our oxen are much inferior, to those of our northern friends."[50]

Although southern observers were at a loss to explain how oxen could outpull horses, they concluded care must be a decisive factor. If oxen were stabled, curried, fed, and cared for like horses, they could work like horses. Writers often cited the inferior treatment oxen received as the primary cause of their inability to work as hard as horses. Farmers may have been more willing to neglect oxen than other draft animals, and as a result the ox's ability as a laborer suffered. One ox supporter argued that "to judge of the capabilities of the ox by the badly-used, houseless, over-tasked, and half-fed animals we sometimes see in the yoke, is doing him great injustice."[51] Most farmers agreed that oxen were the cheapest of the three animals to raise and the easiest to care for, so this may have contributed to their abuse.

Oxen, as a rule, apart from being labeled the slowest of the three animals, also brought the lowest prices. In 1820, one journal reported that an ox cost approximately one-half of what would be paid for a horse.[52] While specific prices for oxen were seldom discussed in the journals, James Shelby of Fayette County, Kentucky, expected to receive about $60 for each of his mules in 1833. By 1850, "[a]nything of a mule" brought between $60 and $100, and "large and fine" mules cost between $100 and $125.[53] Raising steers to working age, usually three years (although many farmers began working oxen at the age of two), cost but one-third as much as caring for horses or mules. Observers believed that oxen could be fed a greater proportion of roughage than horses and remain healthy. Indeed, one Mississippi planter wrote that the greatest waste committed on plantations that used oxen was feeding too much corn to the animals. He blamed the problem on his slaves' ignorance of the oxen's digestive system. Advocating feeding only at night, this owner revealed much about his attitude toward his slaves when he asserted that a "negro can't understand how it is that anything will not do best by eating three meals a day."[54]

Oxen also demonstrated the ability to produce more of what they ate in the useful form of manure. A general rule stated that an ox made about twice the manure that a horse or mule did.[55] Of course, this might have mattered little in the South, where there was typically less interest in manuring fields than in the North. A final advantage of using oxen was that they could be eaten if they grew too old for labor or proved intractable.[56]

Although oxen continued to represent a viable alternative for many southern farmers throughout the antebellum era, mules became the animals of choice in the cotton-, sugar-, and rice-producing regions. Even so, prejudice often stood in the way. If the ox carried the stigma of natural languor, the mule had the reputation of perverse stubbornness to overcome. Mules reportedly had the "natural tendency . . . to be lazy and obstinate."[57] At least one writer felt that there existed a biblical denunciation of mules. In his words, "it would seem a fair inference from the prohibition (Thou shalt not let thy cattle gender with a diverse kind), contained in the nineteenth chapter of Leviticus, that the ugly mule race was much despised or wholly unknown in Judea."[58] It was, therefore, unnatural and contrary to the will of God to produce or use mules. Most southerners must not have agreed with this interpretation of the verse, even though the South as a whole embraced orthodox Christian theology, since mule use spread so widely.[59]

Mules carried the weight of other preconceptions which prevented some farmers from purchasing the hybrids. Detractors labeled the mule a naturally slow creature. Unlike the "patient" ox, critics asserted that the mule simply did not want to work. Many readily admitted the mule's capacity for "steady continued draught, as in threshing, grinding, and other machinery," but few observers believed in its suitability for faster work under saddle or in harness. Advocates, however, refused to accept the mule's slowness as its nature, instead blaming its lethargy on poor breeding or slovenly training. Mules properly bred and trained could serve any purpose, even that of drawing a carriage at the rapid clip of eight miles per hour. A correspondent to the *Southern Cultivator* related that he and two other companions took a carriage ride behind a pair of mules with "Gen. James Shelby, who drives nothing else in his private carriage, from Lexington [Kentucky] out to his magnificent blue grass farm, eight miles, behind two mules of about 15 hands, within an hour, and without the touch of the whip." Shelby informed the writer that the pace was not unusual, for he had driven a pair of mules forty miles in six hours, with a one-hour stop along the way.[60] Shelby was atypical, if not unique, in his use of mules as carriage animals, but since Kentucky was one of the chief mule-producing states, it was in Shelby's interest to demonstrate the adaptability of the animals to varied applications.

Southern planters as a group demonstrated a growing inclination to use mules as draft animals as the nineteenth century unfolded, but northern farmers did not follow suit. Before the mid-eighteenth century, northern farmers usually employed oxen on their lands; after that date horses became common work animals, especially in commercial farming areas.[61] As late as 1821, Benjamin Colman observed of northern farmers that "almost all the draft work is performed by oxen."[62] Still, mules were not unknown north of the Mason-Dixon Line, nor were they without proponents in the nation's more northern regions. Some limited mule breeding for sale occurred in New England during the antebellum years.[63]

A Delaware farmer reported that while mules saw very little use in his area, he expected that they would be employed widely in a few years as farmers came to appreciate the animals more.[64] Pennsylvania coal miners found mules more useful than horses in and around mines and ironworks.[65] A letter to the *Farmers' Cabinet* from Middletown, Pennsylvania, argued in favor of adopting mules for farmwork in the writer's region.[66] Such recommendations swayed few, for Pennsylvanians owned only 2,259 mules in 1850. A decade later the number had grown to only 8,832, a trifling percentage of the more than one-half-million draft animals in the state.[67] Mules impressed even fewer New Englanders, where the total mule population in 1860 stood at 357. The *Agricultural Census* noted accurately, if dryly, that "it is evident that the mules are not favorite working animals in the New England States."[68]

As mule use increased in the South, and more people came in contact with the animals, many masters considered the mule a creature naturally inclined to eccentricity. Among the mule's alleged peculiarities was an inclination to be of a cooler temperament than a horse, a tendency to kick more than a horse, a proclivity to be more gregarious than a horse, and a decided bent for revenge. One farmer warned mule owners to be kind to their charges, "for if treated kindly, they will be gentle; but if treated inhumanly [*sic*], they will treasure up their revenge for years, until an opportunity offers to gratify it."[69]

While mules often seemed determined to make their owners' lives difficult, observers commonly recounted stories relating to mules' affection for one another. Separated from a workmate for a day, one mule, labeled a "perverse beast" by its owner, climbed into the barn's hayloft in its search for an escape route. The writer concluded that it was nearly impossible to confine a mule away from an associate. According to him, a mule would "climb over a fence if practicable like a dog, or if more practicable creep through a crack, or worm himself under it like a pig."[70] While considered by many to be perverse, refractory, stubborn, vicious, unruly, peculiar, and eccentric, mule advocates argued that the animals' detractors misunderstood mules. Mules were not perverse or eccentric, they argued, but more intelligent than horses or oxen.

Ill behavior by mules was sometimes attributed to harsh treatment, not to their nature. Some voices argued that mules were actually more sensitive than horses to the treatment given them. If properly treated, mules would be more willing workers than horses but, if mistreated, would react even more negatively. If the mule "could tell man what he needed most it would be kind treatment," wrote one mule authority.[71] Another observer noted that mules properly cared for were quite willing to work hard. He had often seen a span of mules on the road that were "cheerful as larks, sleek as moles, and handy and nearly as intelligent as human beings." In addition, he noted that they were "never sick, never sorry, but up to everything, even if it be a little mischief at leisure times, such, for instance, as eating up their mangers, or gnawing a hole in the sides of their stalls, when nothing better to do, but always most happy when usefully employed." As proof of their willing natures, the commentator explained that when the two animals were let out of their stalls to water on a Sunday morning, they would "put themselves in their places at their own wagon, which they know from others, and stand ready for action, although without harness."[72] For all of their "immediate intelligence," however, mules were also said to be creatures that needed retraining if not worked for a number of weeks.[73]

Unfortunately for the mule, owners universally believed that the animal could stoically withstand abusive treatment that would render a horse useless. In the context of the slave regime, this was one of the most compelling arguments for mule use. The mule's "faculty for endurance is almost incredible," noted *De Bow's Review*.[74] Mules, another article concluded, "are better for our servants to handle, as they can stand neglect and violent treatment better than the horse." Neither did blemishes, such as the loss of an eye, degrade the value of a mule as much as that of a horse.[75] Supporters insisted that the mule was less likely to chafe or gall, a problem to which the horse was believed to be especially prone, particularly when the animal drove machinery by following a circular motion. Although not universally recommended, many farmers believed that mules could commence field work at two years of age given their tougher physical characteristics, whereas horses were often given another year before being harnessed for heavy work.[76] A lone voice argued that mules should not be worked hard until four years of age, but few mules were afforded that much time to mature.[77]

Not only could mules work hard at an earlier age, advocates asserted that they lived longer than horses. Most writers observed that mules lived an average of thirty years, almost all of which could be spent in harness. One commentator went so far as to calculate that "one mule to a lifespan will kill seven horses."[78] Some fantastic stories circulated which reinforced the belief in the mule's longevity. Two mules, for example, were reported to have reached seventy years of age each.[79] One convert even went so far as to relate through a tortuous logic that he "was once asked, by way of calling my attention to the longevity of mules, if I ever saw a dead mule? I never did. I do not pretend to say they never die, but they certainly live to be very old." He drove home his argument by relating Pliny's account of an eighty-year-old mule, too old to carry a burden, which still followed his coworkers in their daily labors.[80] Even while mules may not have actually lived longer than horses, they carried with them that reputation, perhaps because they could work longer than their counterparts.

Much of the mule's reputation for long life rested upon resistance to disease and injury. "But who ever heard of a ring boned, spavined mule?" asked one editor.[81] That they remained healthy seems surprising when one considers the lack of proper feed and treatment they received. For example, draft animals commonly lived outdoors in the South much of the year, even on plantations. Stables were often little more than crude log structures which offered slight protection from the elements.[82] The mule's reputation for virtual indestructibility undoubtedly contributed to the abuses heaped upon it. In a way, the mule became a victim of its own hardy nimbus. Time and again, correspondents reiterated how ideally suited mules were to the environment of the Lower South.

Perhaps more than the southern environment, ignorance about the most basic aspects of animal husbandry meant that animals had to be strong to survive. The simple fact was that disease was rampant, abuse was common, labor was hard, and only the hardiest of field animals would survive in the antebellum agricultural world.[83] In a lengthy and detailed exchange among farmers within the pages of the *American Agriculturist,* one catches a glimpse of a mule's life, treatment, and death. A Florida planter named Gaston asked his fellow farmers for advice concerning the care of his mules. Many of his four-legged laborers had died, apparently of colic. Gaston lamented that he was among "the most unfortunate people in the world as regards our stock. I lose three, and sometimes five mules every year by the colic; every day there is a mule brought to me from the field with colic. Now I cannot see the reason of this, unless it is our mode of treating them." Gaston's method of caring for his animals was indeed questionable. It included plowing the mules hard and then giving them "as much water as they can drink." In addition, they received free-choice corn in a trough at the same time. It was a recipe for disaster, even for the usually prudent and hardy mule, and free-choice feeding remained common over the decades.[84]

The diet and treatment of mules reflected the lack of knowledge about animal husbandry, as well as the subjective nature of observations about what mules needed. Contrary to what many mule owners suggested the animals be fed, James Boyle warned that feeding wheat or rye straw would render a mule unfit for hard service. He recommended "crushed corn and cobs" in addition to roughage, and not cornmeal. He explained that "mules are but little subject to disease, except by inflammation of the intestines, caused by the . . . excessive drinking of cold water after severe labor, and while in a state of high perspiration."[85] The *American Agriculturist* asserted that if one had to give very cold water to hot mules, "a handful of hot wood-ashes" thrown into the bucket would prevent any ill effects such as founder or colic.[86] Although less susceptible to health problems than horses, as was indicated by studies done during World War I, mules were hardly indestructible and needed a modicum of care to stay fit for work.[87] Unfortunately, many mule caretakers lacked the knowledge, desire, or both to keep their charges in good health.

M. W. Phillips responded to Gaston that "no man has any right to accuse our Maker of partiality, who will treat stock" as did Gaston. Phillips's mules received royal treatment in comparison to Gaston's. His teams began the day at four o'clock in the morning and worked until eleven. During the midday break, the mules wallowed and rested in a lot until cool. Then the servants watered, fed, rubbed down, and curried their charges. Work began again at two o'clock, lasting until dark, when the cooling, watering, and feeding process began again.[88] Some planters rested their mules on Sunday, requiring slaves to walk if they wished to visit another plantation or church. James Bolton noted that while the slaves on Whitfield Bolton's estate never worked on Sunday, they had to walk the nine miles to the nearest church since "Marster diden' 'low his mules used none on Sunday."[89]

Even concerned owners with the best of intentions might offer cures that hindered recovery or caused greater injury and pain to the sick animal. Responding to a mule's illness diagnosed simply as "in his head," John Horry Dent had his charge "well smoked with tar, [and then] split the skin of his forehead and put mustard in the incision." Dent commented incredulously that "nothing seems to relieve him." The mule survived the treatment but for a day.[90] An authority on mules confidently exclaimed that "[b]leeding at the mouth will cure them of nearly every disease; and by being turned out on pasture, will recover from almost any accident."[91] Even as discussions of mule diseases and ill-conceived cures took place in the journals, more southerners switched to the animals, largely because of their alleged indestructibility.

Mules were reported to have the ability to subsist on both less feed than the horse as well as inferior rations. This could only endear the animal to planters in a region where good pasture was the exception and where cash, not food, crops were the focus of many plantations. One mule owner reported, in what would become a commonly related mule story with numerous variants, that he had discovered how little his mules needed in the way of feed by accident. Filling the animals' hayrack with wheat straw one night, he was surprised to find that his mules had consumed all of it by morning, even choosing it over their corn fodder. Each night thereafter, he reduced the fodder and replaced it with a proportionate amount of wheat straw. Ultimately, he switched to the less costly wheat straw entirely. He argued that this roughage, along with the usual ration of corn or hominy, sustained his animals, although he worked them hard all winter. This article went on to assert that widespread adoption of mules could have far-reaching effects. Should the Irish nobility shift to mules, the food shortage in their land might be averted, and the "grave might be cheated out of thousands of its victims who starve to death for want of grain that horses consume."[92]

To some owners, the mule's quirky palate was a source of amazement, especially since the mule was often regarded as more intelligent than the ox or horse. One writer noted that if "a mule be bedded with the commonest salt hay, and his manger filled with good oats, underlaid with a half peck of thistles, he will probably eat the thistles first, his bedding next, and the oats afterwards, unless intermediately he should take a notion to feed on his crib or the side planking of his stall."[93] Another benefit that mules provided owners, later substantiated by formal studies, was that their digestive systems proved far less delicate and therefore not as prone to upset as those of horses. Colic and founder, then, were less common in mules than in horses, because the mule's digestive processes are less affected by changes in feeds and feeding, and because mules tend not to gorge themselves on feed or water like horses.[94]

Partly as a result of their alleged penchant for inferior feeds, mules acquired the reputation of being cheaper than horses to raise and keep. "The horse is costly and extravagant—the mule cheap and economical," concluded one correspondent to the *Southern Cultivator*, adding that he believed mules required but two-thirds of the feed consumed by horses. The same writer reported that the Farmers' Society of Barnwell District, South Carolina, had concluded that two mules could be raised at less expense than one horse.[95] Another farmer calculated the relative costs of owning a team of ten horses versus a team of ten mules over a twenty-year span. The analyst estimated that the horses and mules would consume an equal amount of hay, but that the horses would eat twice as much corn as the mules. He predicted that the horses would cost $6,600 over two decades, including feed and shoeing costs. The mules were not thought to need shoeing. Overall, their care would cost $3,000 over the same period, resulting in substantial savings.[96] There would also be corn left over which could be marketed to the benefit of the producer and consumer.[97]

While mules were believed to be cheaper to raise and keep, they were not less expensive to purchase. Mules consistently cost more than workhorses or oxen due to a number of factors ranging from the need for a jack to stand at stud, to transportation costs from the mule-raising sections of the nation, to increasing demand in the cotton South. The United States Patent Office printed a North Carolinian's report in 1853 that mule prices ranged from $100 to $150, while "good" horses sold for $100 to $125.[98] Although prices fluctuated locally and varied widely from state to state, "good" mules averaged between $100 and $150 in cost late in the antebellum era, while ordinary ones brought approximately $70, although many variables influenced prices.[99] The higher cost of mules may in part explain why planters, who had more resources than farmers, would be among the first to experiment with and adopt mules.

Size and quality influenced prices as well as demand, but mule prices appear to have remained fairly consistent during the last decades before the Civil War, despite complaints from buyers like John Horry Dent, a successful "planter-farmer" whose agricultural career spanned the middle and late nineteenth century in South Carolina, Alabama, and Georgia. Dent recorded in his journal in 1856 that he had purchased two mules "for $245—125 for the grey [and] 120 for the sorrel," adding that at these prices he considered mules "exorbitantly high—higher than I ever knew them before."[100] Three years later Dent purchased a mule for $140 and considered the transaction "a bargain."[101] In early 1860 Dent purchased two more mules for $350, although he does not state the price for each one.[102] In Georgia high-quality mules sold for approximately $300 per pair in 1855.[103] On the eve of the Civil War, the *Southern Planter* pointed out what had become a southern truism, that the "ordinary cost of a fine mule is much greater than that of a good farm horse."[104] Higher costs for mules remained a fact throughout the era of mule use in the South, but cost did not prevent their adoption.[105]

Southerners did not all adopt mule agriculture equally, and the distribution of the different types of draft animals was not uniform in the region or in individual states. Southerners tended to connect mules and slaves in their minds, and tied both to the plantation system of the Deep South. A Georgian commented that "in no part of the country is the mule better adapted to all the purposes of husbandry for which the horse is used, than in our own State."[106]

Mules, slavery, cotton, and large plantations formed a southern constellation in fact as well as in the popular mind. During the late antebellum period, only one-fourth of southern farms without slaves employed mules. More than sixty percent of farms with fewer than fifteen slaves used mules, while more than ninety percent of medium and large plantations used mules. In addition, the fewer acres and slaves a farmer owned, the more likely he was to rely upon a single type of draft animal, usually the horse or ox.[107] Paradoxically, at the same time that mules became identified with plantations, they were less likely to be the only type of draft animal used on large estates.

Large planters could afford to organize their operations in order to employ oxen for heavy work, mules for plowing and cultivating, and horses for pulling carriages and riding chores.[108] John Horry Dent employed different types of animals in his farming operations. He recorded in his journal, for example, that his oxen were being used during February hauling manure and cottonseed to the fields and firewood to the house. Dent also owned mules but apparently did not as a rule use them to haul such loads.[109] Numbered among the animals on Oktibbeha County, Mississippi, planter John Washington Rice's estate in 1860 were one yoke of oxen; five horses, three of which were mares; and twenty-four mules. Rice exemplified the large plantation owner who used draft animals for specific duties meted out to his sixty-eight slaves.[110]

Because only large planters could afford such specialization, almost half of southern farms that did not employ slave labor relied on but one type of draft animal. Approximately ten percent of free farms employed all three types of animals. As plantation size and the number of slaves increased, so did the likelihood that more than one draft-animal type would be used. On farms with fewer than fifteen slaves, fully one-third had horses, mules, and oxen, while twenty percent used only one type of animal, usually the horse. Seventy percent of the plantations that housed sixteen or more slaves had all three types of draft animals, and more than ninety percent used at least horses and mules.[111]

Since mules and plantations were closely related, the identification of mules with the plantation system reflects more than a curious aspect of southern antebellum society. The decisions planters made as individuals and as a class on matters important and mundane, such as deciding which type of draft animal to employ in the field, reflected the intersection of cultural, economic, and political considerations. The type of draft animal a planter chose to purchase or raise for use by his bondsmen reveals a glimpse of the planter as a person. The same point applies, of course, to southern yeomen who farmed within a specific context and whose decisions about plowing and planting were influenced by local conditions and individual considerations.

Although mule use expanded rapidly in the decade before the Civil War, all three draft-animal types were common in southern barns and fields in 1860, revealing a little-noted aspect of the complex antebellum social structure. It may be that while yeomen and planters were both masters of sorts, they were masters of different animals.[112] Because mules, slaves, and planters shared ties, mule ownership may reveal important points about particular segments of southern society, especially since the antebellum South was not immune to social tensions.[113] Planters, to some degree imbued with the American focus on wealth and economic success, participated in a market economy by producing staples, selling them on the world market, and using those proceeds to purchase goods produced by northern industry.[114]

While planters negotiated with factors in Charleston, New Orleans, New York, or London, however, non-slaveholders participated in a much more local economic universe. Thus, some differences existed in how slaveholders and non-slaveholders viewed the world.[115] This is not to say, of course, that major rifts existed in the social structure of the South, but that, to a degree, the planters and yeomen made choices based on different needs and desires.[116]

Draft-animal use may be one important means of highlighting differences between those satisfied with their status as small slaveholders or non-slaveholders and those aspiring to enter the ranks of the plantation class. For example, a correspondent to the *Southern Planter* revealed significant regional variations in the use of draft animals when he wrote, "I suppose you, with your tide-water prejudices, are an advocate of mules for plantation use."[117] While the ratio of mules, horses, and oxen varied widely within Virginia's border, mules generally played a larger role in eastern Virginia than they did in the hilly or mountainous areas of the state.[118] Likewise in Texas, there were significant regional differences on the eve of the Civil War in terms of mule versus horse use, and these differences highlighted more than a simple preference for one draft animal over another.[119]

Another writer to the *Southern Planter* noted this regional theme in 1852 by expressing surprise that relatively few farmers had switched to mules. He explained that an improving transportation system affected the number of mules in a given area. Without the means to export surplus corn easily, farmers found it more economical to feed their grain to horses, which consumed more of the crop than mules. Thus, farmers in isolated regions retained the more gluttonous horse because he ate more of their corn. But when it became easier to market corn, farmers preferred selling their surplus; mules, being lighter eaters, left the farmers with more grain. Having "become accustomed to the horse and this wasteful mode of feeding him, our farmers have come to regard it as all right and proper," stated an observer. But, he continued, with the opening of more lucrative markets for corn generated by better roads, steamships, and railroads, the "old, extravagant and wasteful habits" were passing away.[120] Improved transportation also allowed mules to be brought to an area more easily. It appears the more closely and completely an area was tied to the market economy, the more prevalent mule use became. Thus, economic development and the adoption of mules were closely related in the antebellum South.

Mule use expanded with the slave and cotton empire. As planters accepted the superiority of mules for their own uses, especially in the cotton-growing regions, mules became more and more common in fields worked by slaves.[121] Along Georgia's upper Savannah River, mule ownership proved the exception during the early nineteenth century, but by 1860 Elbert County had more than eight hundred mules, nearly twenty-eight percent of the county's draft animals.[122] But in areas where slavery and cotton did not predominate, mules played a less important role than horses or oxen. The reasons planters shifted to mules appear relatively clear. Planters who chose to use mules were convinced that under plantation conditions the animals were superior to alternatives. At the same time, however, it is at least as important to question why some southerners did not shift to mules.

Several South Carolina districts exemplify the differences in mule distribution. In the upstate districts of Anderson and Pickens, mules made up less than one-fourth of the total number of draft animals in 1860. In 1850, whites made up two-thirds of the population in the upcountry region which contained these districts. This area of the state boasted few plantations, while small and medium farms dominated the landscape.[123] South Carolina's lower piedmont districts, on the other hand, relied more heavily upon mules. In 1860 mules made up forty and sixty percent, respectively, of the total number of draft animals in Edgefield and Fairfield districts.[124] The lower piedmont had a larger black population than the upper piedmont and a plantation economy.[125] Oxen played a greater role in upstate agriculture than they did in plantation areas as well. In Anderson and Pickens districts, for example, working oxen made up eighteen and twenty-eight percent, respectively, of the total number of draft animals used in 1860. In Edgefield and Fairfield districts, however, oxen comprised only thirteen and five percent, respectively, of the 1860 draft-animal population. In

Pickens district, there were nearly twice as many oxen as mules in 1860.[126] In many areas of South Carolina, then, mules were not the primary work animal on farms on the eve of the Civil War.

The same general observations hold true for Georgia in 1860. In the Blue Ridge counties of Towns and Habersham, mules made up less than fifteen percent of the work-animal population. In the piedmont region of the state, the percentage of mules was significantly higher than in the mountain areas. For example, in Baldwin and Greene counties, mules made up forty-one percent and thirty-four percent of the draft-animal populations, respectively. The percentages of mules in the Central Cotton Belt counties were higher still. In Houston County, mules made up more than fifty-five percent of the draft animals, while in Twiggs County the mule percentage was nearly forty-seven percent.[127] In short, regarding draft-animal usage, important distinctions between counties and regions, as well as within them, yet to be fully explored, existed in the Old South.[128] What is clear is the difference between areas of small and medium-size farms and plantation regions. Since mules represented agricultural innovation, the plantation South, that is the South most heavily dependent upon slavery, led the yeoman-dominated areas in adopting at least one progressive agricultural technique because it strengthened slavery's footings. No wonder the agricultural journals echoed with optimism.

Still, the differences between upcountry yeomen and low-country planters were of degrees. While the two groups did not hold fundamentally different worldviews, they did live in distinct, if overlapping, social and economic universes.[129] The same factors that influenced planters to devote larger percentages of their lands to cotton than smaller farmers often led to differing choices of draft animals.[130] Abandoning a safety-first agriculture by investing in unnecessary or excessively costly animals could overextend a farmer's limited capital. On the other hand, if small farmers employed hired slaves, or aspired to slave ownership, it might make sound economic sense to purchase mules. Some southerners did move into and out of slaveholding ranks, but it is difficult to determine its scope. Since the link between slaveholding and mules is manifest, mule ownership by farmers might also reveal an individual's inclination toward the slave regime, his aspiration to join slaveholding ranks, or his recent fall from that position.[131]

As the South wrestled with the issue of the nature of the Union and its place in the nation's future amid the turmoil of the 1830s, with slavery and plantation agriculture at stake, its leaders were concerned not only with political and philosophical issues, but with the viability of southern culture. Slavery had to be shown as adaptable. The mule provided a means to do that. Ironically, the desire to defend slavery prompted southern planters to search for suitable innovations. The adoption of the mule, as well as the debate surrounding its applicability to southern conditions, illuminates the ability of some planters to champion and embrace reform, albeit reform tempered by an overpowering urge to sustain and strengthen the status quo, even as their thoughts on the slavery issue ossified and sectional tensions led to armed conflict.

While it is impossible to determine whether mules would have come to dominate as the main draft animal if the Civil War had not occurred, it is clear that in the years after the war, mule use increased in the upcountry and lowcountry South. Whereas mule use had been uneven in 1860, mule use in the postbellum years became almost universal. Horses and oxen did not disappear, but the mule rose to preeminence on southern farms and plantations. This is in some part ironic because of the impact that the war had on the number of mules in the nation.

The Civil War decimated the southern mule population, although how many mules died as a result of the war is difficult to ascertain. Northern and southern war efforts absorbed enormous numbers of the animals, as well as disrupting normal breeding and trading routines. The mule population of some states was reduced more than in others. Georgia, for example, lost sixteen percent of its mules, while Alabama and Louisiana lost thirty-two percent of the same. Harder hit still were Virginia and Arkansas. Each of these states lost nearly forty percent of its mules.[132]

Because of their toughness, mules tended to fare better than horses or cattle in peace and war. Thus, in many areas, mules increased in proportion to horses and oxen during the war. As soon as hostilities ceased, the mule renewed its climb to the dominant position in the South's agriculture, also aided by the fact that thousands of Confederate soldiers had discovered that the mule's peculiar attributes made it a good choice as a southern work animal.[133] The war, then, acted as a catalyst of agricultural change in the South by disrupting the agricultural status quo and introducing many farmers to mules.

The connection between mules and blacks, rooted in the antebellum southern experience, only grew stronger as sharecropping spread in the late nineteenth century. Just as slaveholding planters had dictated which animals would be used on their acres, postbellum landlords generally saw to it that their lands were plowed with mules. With so much change occurring in the South, landholders may have been more apt to cleave to whatever traditions they could. As a result of increased mule use, the mule began to achieve its premier place in the southern draft-animal pantheon between 1870 and 1880. By 1880 the South had 300,000 more mules than in 1860. By 1900, mules were fairly evenly spread over the cotton South. Areas that had been marked with a low density of mules changed as more mules were hitched to plows on southern farms throughout the region.[134]

By the twentieth century, mules were a standard fixture on southern cotton farms and plantations of all sizes. The number of mules in the region increased until 1925, when nearly four-fifths of the mules in the United States were in the South.[135] Within the agricultural realm of draft animals, it appears as though little changed after the turn of the century. Improved farming practices and better implements made their way onto many southern farms, but by the same token, many farmers continued to use hopelessly worn-out tools and highly damaging cropping techniques. But the mule remained a constant, as well as a central, aspect of the southern agricultural identity. This is evidenced by the fact that farm size in the South was often based on the measure of a one-mule farm or a two-mule farm.[136] Despite the fact that the cotton South organized its agriculture around the animals, the region produced few of its own mules. As a result, the breeding and trading network that supplied farms and plantations with mules was a crucial component of the southern agricultural system.

2

Supplying the South

Mule Breeding and Mule Trading

Southern farmers depended heavily on mule breeders and a mule-market network to supply the vast majority of the animals to the South. The extensive mule-breeding industry and the mule trade expanded during the late eighteenth century as cotton swept the mule along its course across the southern landscape. Border states like Kentucky and Tennessee developed mule breeding and mule trading with the cotton South originally, but were joined after the Civil War by more western states such as Missouri and Texas.

Millions of mules streamed into the Deep South from mule-raising states to replace worn-out animals and to expand cotton production. Mule dealers roamed the South, providing the link between producers and customers, and on the local level mule traders also played a role. A distinctive mule-trading culture developed in the process. The decline of the mule-breeding industry and the mule trade spelled the end of mule dealers and traders, and they, like the animals they bought and sold for a century, disappeared from the southern landscape—both man and creature victims of post–World War II technological change. In the end, the mule dealer succumbed to the tractor dealer, just as the animal he sold gave way to the machine.

At the outset of a discussion of mule breeding it is important to note that, no matter where mule production takes place, ending up with a mule is not a simple affair. For example, many jacks will refuse to mate with mares after coupling with their own kind, that is jennets or jennies. Thus, some jacks and jennets may be used solely to continue the line of jackasses. Other jacks may be used only to breed with mares to produce mules. Finally, stallions and mares must be mated to produce the mares for crossing with the jacks.[1] Horses and oxen could be produced locally and purchased from a neighbor, or bred on one's farm with less trouble and expense than mules required.[2] Producing mules is, in short, a complicated business, and this helps explain why southerners usually purchased mules from outside sources.

The southern states gained a reputation for producing low-quality horses and mules. François André Michaux traveled in the United States in the early nineteenth century. He observed that southerners gave little consideration to improving their plow horses. Kentucky's horses were small with a "bad appearance," but they were "still worse in Georgia and the Upper Carolinas."[3] Southerners apparently possessed the same lackadaisical attitude toward improving mule stock. Harvey Riley described the mules of eastern and southern Maryland as "shabby." He stated he had never seen "more broken-hearted, poverty-stricken, and dejected-looking" teams anywhere. While he admitted Maryland produced good horses, he called for the state's inhabitants to produce "a better kind of mule," but it would be nearly a century before high-quality mules raised in the cotton South were more than a curiosity.[4]

Mules were unusual in the English colonies and the new nation, but the extent of mule production in the colonies is unknown.[5] Before the Revolutionary War, it is likely that few, if any, of the small, inferior mules were bred in the colonies. Most were English mules, imported from the Hampshire area.[6] A few mules came from other sources. The *Southern Planter* declared that prior to 1783, "scarcely any mules were to be found in the American Confederation; a few had been imported from the West Indies, but they were of diminutive size and of little value."[7] A leading authority on southern mules found but a "thin and scattered distribution" of the animals until late in the antebellum era.[8]

Horses and oxen were by far the more common work animals from the early days of settlement.[9] The slight cultural influence of the Spanish throughout most of the English colonies, as well as a dearth of knowledge regarding mule breeding and mule raising, limited the use of mules during the colonial era. Colonists from East Anglian England, a large segment of the first wave of English settlers, came from an area where asses were plentiful, but they did not bring any with them, nor did Dutch and German immigrants.[10] Thomas Jefferson, interested in many aspects of improved agriculture, used all three types of draft animals on his estate but preferred mules over horses and oxen. Known for his wide-ranging interests and agricultural experimentation, Jefferson apparently did not attempt to carry out a mule-breeding program, instead choosing to purchase the hybrids employed on his acreage.[11]

Mule breeding held more interest for George Washington, the leading mule advocate of his day and a crucial figure in the advancement of mule use in the southern regions of the United States. While it is unclear when Washington became enamored with mules, he came to believe mules to be superior draft animals and worked diligently to promote their use and to inaugurate a successful mule-breeding program at Mount Vernon. Like Jefferson, Washington experimented with much in agriculture, but in terms of breeding, he sought in particular to improve his sheep and mules.[12] One antebellum journal asserted that in the agricultural realm, Washington's advocacy for mule use stood as his most important contribution.[13] The tale of Washington and mule breeding illuminates the difficulty anyone interested in breeding mules faced in the late eighteenth century.

Because of his national and international stature, Washington probably accomplished much in that vein that would have been impossible for men of lesser repute. Washington hoped to change what he considered the wasteful character of American agriculture by developing the foundation stock for an American mule industry. Washington's correspondence related to jacks and mules reveals an intense interest in the animals, but disappointment that he could spend so little time personally overseeing the jack- and mule-breeding program at his estate. A "good jack would be a public benefit to this part of the country, as well as a private convenience to myself," he wrote in 1784.[14] Reflecting what would become a standard assumption about the animals, Washington believed mules to be ideally suited to southern agriculture particularly because of their ability to withstand "the care which is usually bestowed on draught cattle by our Negroes."[15]

Spain held the crown as the world's leading breeder of jacks, and it was there that Washington turned. Washington desired not only first-quality jacks, but those with proven procreative potential, for he held suspicions that the Spanish, jealous of their virtual monopoly in quality jacks, exported no jacks capable of siring offspring. Washington suspected that those jacks considered unfit for breeding and those intended for export by the Spanish "very frequently have their generative parts so injured by squeezing, as to render them as unfit for the purpose of begetting colts, as castration would, when from a superficial view no imperfection appears." He added that he was not certain the Spanish employed such a technique, but that he believed the possibility of such mendacity warranted notice.[16]

Despite his misgivings concerning the fecundity of foreign jacks, Washington obtained shipment of four jacks and two jennets to America, two jacks from Spain's Charles III and the others from the Marquis de Lafayette. Of this group, two jacks perished en route. Washington dubbed Charles III's jack Royal Gift and Lafayette's the Knight of Malta. The two jacks served many mares, but perhaps more importantly, Washington bred Royal Gift to a jenny which resulted in an Andalusian-Maltese jack named Compound, one of Washington's favorite jacks.

Although Washington's breeding program promoted the use of mules as far as Charleston, South Carolina, where Royal Gift spent three breeding seasons, he was disappointed with the impact of his vision of populating the nation with mules. Royal Gift's performance during his southern tour did not satisfy many, and may have engendered some prejudice among the planters whose mares he covered but allegedly did not fertilize. Washington became president during the formative years of his mule-breeding program, and public service limited the amount of supervision he could provide. Still, at the end of Washington's two terms, forty-two mules pulled plows and carriages on Mount Vernon lands and an indeterminate number of his jacks' offspring worked on southern farms.[17]

While Washington's stature could only have added to that of mules, personal endorsement was not enough of a reason for most farmers to switch to mules. Indeed, Washington's state of Virginia never adopted mules to the extent that many southern states did. It is impossible, then, to judge accurately Washington's true impact on the adoption and breeding of mules in the United States, except to note that his actions stimulated both mule production and mule use, especially in the South.[18] At the turn of the nineteenth century, however, mule use was far from universal in any area of the nation. Not until the 1820s did mules appear in any consistent fashion in the agricultural press, and then they were at first treated as a curiosity as much as a viable alternative to horses or oxen.

Like George Washington, Henry Clay played an important role in advocating mule use. As in Washington's case, Henry Clay's impact on the shift to mules is difficult to assess, but he achieved nearly mythical status among mule raisers. Doubtless, Clay's political stature, like Washington's, added weight to his support of mules. Whether Clay led the South to adopt mules or simply rode an already swelling crest of mule popularity is unclear, but for more than twenty years Clay actively sought to improve American mule stock and to an appreciable extent laid the foundation for Kentucky's highly successful antebellum mule industry. At the least, Clay's advocacy contributed to the rise of mule breeding and mule use in the South.[19]

Without an increasing desire on the part of southern planters to innovate, neither Washington's nor Clay's advocacy would have succeeded in convincing southern agriculturists to purchase and use mules. It is no coincidence that Clay, the eloquent proponent of the forward-looking American System which so clearly reflected an optimistic vision of a technologically advanced, market-oriented nation, became one of the best-known champions of mules from the late 1820s.

Henry Clay had a deep interest in stock breeding of all types and firmly believed in the utility of blooded or purebred stock.[20] Although he may have achieved recognition as a horse breeder, and his more important role as a livestock breeder may have been tied to cattle, it is safe to say that Clay was crucial to the growth of the mule industry in the United States.[21] The garrulous Clay often escorted visitors to his Ashland estate around his grounds, fields, and stables while expounding upon the virtues of his stock and crops.[22] Clay's correspondence during his extended absences from Ashland also reveals an intense interest in the animals at his estate. While Clay had broad stock-breeding and agricultural interests, jacks, jennets, and mules retained a prominent place in his farmer's heart if his letters are any indication. In 1831, Clay owned approximately eight hundred acres, fifteen "hands," nearly one hundred head of horses and mules, and about one hundred head of cattle, in addition to numerous other animals.[23]

Clay's most important contribution to Kentucky's and the nation's mule industry was his importation of high-quality breeding stock, which laid the foundation for a unique American mule by "greatly improving" American jack stock.[24] When Clay decided to raise mules is unknown, but it is clear he was not the first in the Kentucky Bluegrass to do so. In 1827 Clay's overseer, John H. Kerr, informed him that someone in the neighborhood had a "fine young jack" for sale. Kerr explained that he mentioned the animal because he believed "it would be good policy to have one on the farm."[25] Although it is unclear when Clay first brought jacks to his Ashland estate, mules worked in the fields in the 1820s. Because of his distressed finances, Clay held an auction at Ashland on June 24 and 25, 1825. Included in the sale were household furnishings, horses, mules, cattle, and sheep; jacks were not mentioned.[26] Had Clay owned jacks at that time, it is likely they too would have been enumerated, unless he prized them too much to put them on the market.

Clay's beloved Kentucky Bluegrass region had long been the home to high-quality animals of all sorts, and the state quickly gained a reputation for its excellent livestock.[27] In 1831 Clay observed that Bluegrass farmers raised "vast numbers" of mules for the southern market.[28] But he is credited with the first "valuable importations" of jacks into the region.[29] Clay had experience in importing blooded stock from Europe, including sheep, hogs, cattle, horses, jacks, and jennets, so he understood the risks and potential rewards involved in procuring the animals, having them shipped across the Atlantic, and then brought inland on foot.[30] Shipping could add considerably to the cost of a jack, sometimes totaling more than the purchase price of the animal itself. Clay estimated the price of a Maltese jack in the late 1830s to be $250, but shipping and importation costs could add as much as $300 or $400 to the initial cost.[31]

There were safer and less expensive investments, but mule raising could prove lucrative. Clay benefited from the mule boom of the 1830s and 1840s. Periodic sales of the increasingly popular animals more than once provided a much-needed financial boost to Clay's sometimes precarious economic fortunes. For example, Clay sold twenty-four two-year-old mules to North Carolinian Ebenezer Pettigrew for $2,160. In 1845 Clay sold fifty-two mules for $6,185 and expected to sell ten more at $100 each.[32] Nor was Clay alone in profiting from mule breeding and sales. In the early 1850s mule sales made Kentucky farmers more money than any other endeavor, and mules were often considered the most profitable stock that could be raised in the Bluegrass.[33]

Clay's impact on the American mule dates to the late 1820s when he imported a Maltese jennet and jack. Calypso, allegedly the largest jennet in Kentucky, and Ulysses served well, the former producing seven foals in as many years and the latter standing at stud at Ashland.[34] Clay considered his acquisition of the two animals a coup of sorts, since judging horseflesh always carried some danger. Perhaps suggestive of where his heart truly lay, Clay traded a Washington, D.C., lot "fronting a public Square" for the two animals and any offspring they might produce.[35] Two years after the trade, Clay commented that Calypso and Ulysses had "proved very fine. Both are large, young, well formed, of good color and excellent breeders."[36] In 1829 the jack Achilles was brought to stand at Clay's estate.[37] In the 1830s, the bloodlines of two of Clay's best jacks, Warrior and Mammoth, were combined and a distinct breed, the Kentucky jack, came into being.[38] This combination produced animals that became the foundation for Kentucky's mule production.[39]

In his endeavors to procure first-rate jacks and jennets, Clay purchased wisely and took a close interest in their welfare, partly because of their high cost. In one buying spree, Clay purchased a single jack for $1,000 and a group of four jacks and a jennet for $6,000.[40] From Washington, he wrote to his son, "I am very anxious about all my Stock, but especially in respect to my Asses. The young Jack was very promising when I left home, and I wish him to be kept in good plight, as well as the Jennies."[41] Henry Clay Jr. assured his father that his stock, particularly the jacks, were in good health and being carefully tended.[42] Magnum Bonum, one of the jacks produced at Ashland, so won the statesman's heart that he commissioned the famed animal artist Edward Troye to capture his likeness on canvas.[43]

By the 1830s, perhaps because of Clay's and others' successes and partly because the gospel of improved agricultural practices had begun to take hold in some areas, numerous breeders were bringing blooded stock into the United States. Before the Civil War, numerous outstanding mule and jack breeders resided in the Kentucky Bluegrass region. Among the premier jacks that lived in the area were Mammoth and Buena Vista, the former being particularly noteworthy because of the improvement he visited upon jack blood through his large size and excellent bone.[44] J. R. Brockett imported Mammoth from Catalonia, Spain, via Charleston, South Carolina, in 1819.[45]

A correspondent urged Clay in 1836 to accept any good offer for the jacks he had placed on the block because the writer knew of several ships en route to Malta from New Orleans which were to return with jacks. "Another year," he warned, "[and] you will no doubt be inundated with them."[46] A jack glut did not materialize, however, as southerners swept westward and opened vast new areas of cotton lands which demanded mules from the Upper South. Evidently, Clay's mule trading ended in 1850, two years before his death, just as the mule's explosive increase began, when he sold a group of mules in Alabama. By that date, mule raising ranked among the top of livestock-raising endeavors in Kentucky.[47] By the time of the Civil War, the Kentucky mule had improved, "under the influence of abundant food and a suitable climate, with judicious care and skill in their breeding," to the point that they were equal to any mules in any part of the world.[48]

During the antebellum era, most buyers agreed that Kentucky produced the best mules in the nation. Size was one of the hallmarks of the Kentucky mule, and that reputation served them well in the marketplace. Many planters liked big mules, and many looked no further than height and weight when purchasing a mule. Knowledgeable mule breeders often mentioned size when discussing their jack or mule stock. When Henry Clay purchased an expensive jack in 1833 he singled out size for mention, noting that Warrior was "by far" the largest jack he had ever seen. To Clay's pleasure, Warrior had also proven himself capable of passing on his size by siring at least one large colt.[49] The *Southern Cultivator* recommended that "large and spirited mares" be bred to the largest jacks to produce the ideal mule.[50] By the late 1830s, mules on display at Lexington stock shows commonly reached seventeen hands.[51] Size, regardless of conformation or spirit, became an important measure of a mule's worth in the South.[52] This near obsession with jack and mule size doubtless grew out of a reaction to the preponderance of "half grown, rough looking dwarf mules" which worked on many southern farms.[53]

Some voices counseled moderation on the size issue, however. In a lengthy letter printed by both the *Southern Cultivator* and *Southern Planter,* Mark

Cockerill of Nashville, Tennessee, questioned the value of sheer bulk in a work animal. He was intimately acquainted with stock raising since he owned a five-thousand-acre farm and maintained more than two thousand sheep and seven hundred horses, mules, and cattle.[54] In his estimation, it was not the large mule that the South needed. The "medium sized mule, full of spirit and action" was the ideal animal for "all the labor of the South." He pointed out that "over-sized mules, as over-sized horses, do not wear well." Compared to Tennessee's fifteen-hand, eight-hundred- to nine-hundred-pound mules, Kentucky's large mules could not work well because they were grown to a great size by their owners through a "sort of forcing, hot house treatment." According to Cockerill, Kentucky breeders failed to make a "proper distinction between animals intended for active labor and those intended for the slaughter pen" and ultimately for the table.

Cockerill further explained that the desire for size above all else had driven Kentucky's breeders to produce animals devoid of "symmetry, spirit, action, lasting endurance and permanent value." It was obvious, or at least Cockerill hoped it to be so, that any intelligent farmer would choose the animal that "conforms to the laws of nature," rather than one that violated them. The desirable animals to which Cockerill referred were, of course, Tennessee mules. Cockerill argued that Tennessee's mule producers had not become obsessed with size, but focused on quality. Cockerill did not place all the blame on Kentucky's breeders, however. He acknowledged that uninformed southern farmers demanded large mules and paid in proportion to size.[55] Therefore, his diatribes against Kentucky's large mules sought to educate southern readers.

Stock farmers in Kentucky and Tennessee naturally benefited as mules replaced the other types of draft animals throughout the South. The fact that most high-quality jacks resided in Kentucky was an important factor in that state's early domination of the trade. Henry Clay wrote to a friend that he believed that "there are more good Asses now in Kentucky than in any other Country in the world. The price of good Jacks varies from $750 to $2,000. One that I introduced into this State a few years ago has sold as high as $5,000."[56] By the time Clay penned this in 1837, mules were becoming common on plantations in the cotton regions of the South, and the mule trade was an integral component of Upper South to Lower South trade which included the movement of grains, hemp for cotton baling, and mules.[57] By 1838, Kentucky was sending seven thousand horses and mules to the cotton South.[58] Thus, as the mule trade increased, it did not have to open new relationships or trails, since Upper South traders had long trekked south.

François Michaux observed a lively horse trade between the Kentucky Bluegrass and low-country South Carolina during the early years of the nineteenth century. Drovers took groups of fifteen to thirty horses from Lexington, Kentucky, to Charleston, South Carolina, in eighteen to twenty days.[59] As planters demanded more mules, the equine herds headed south changed in composition and frequency to answer the market. The relationship between Kentucky and South Carolina was not always amiable, however. During the 1828 tariff controversy, Henry Clay's pro-tariff position prompted a boycott of Kentucky mules by angry South Carolinians. In Camden, South Carolina, the local paper gleefully reported that a Kentucky drover "with 30 or 40 very fine horses and excellent mules" remained in Statesburg, South Carolina, ten days without selling even one of his charges, "such is the spirit of independence and opposition to unconstitutional and inexpedient legislation." The ill feelings passed in time, however, and the mule trade soon returned to its normal liveliness.[60]

By approximately 1814 Kentucky had secured markets in Maryland, Virginia, North Carolina, and South Carolina. During the following years, after Creek power had been broken in the old Southwest, Georgia, Alabama, and Mississippi emerged as ready markets for Kentucky pork, mules, and horses.[61] Much of the trade destined for the eastern seaboard states went through Richmond, Virginia, and Knoxville, Tennessee, while the trade bound for the newer cotton states proceeded through Nashville, Tennessee.[62] The importance of different cities as mule markets changed during the mule era in the South. For example, Richmond, Virginia, was one of the largest horse and mule markets in the South during the 1890s. During the 1893–94 marketing season, between eight thousand and ten thousand horses and mules were sold in the city.[63] Around 1900, Memphis, Tennessee, boasted that it was the largest mule market in the South and one of the largest in the nation. In late May 1906, one hundred mules were sold to a levee contractor for $250 each, one of the largest mule transactions to that date.[64]

Mules not large enough for the southern market, or inferior in some other way, often ended up in Baltimore, Maryland, for shipment to the West Indies. Some mules also were sold to work in Mexico and others to labor in eastern coal mines.[65] A few "small and inferior" mules, labeled "deadheads," could not be sold and remained in service on farms in their native regions.[66] Most mules did end up in the South, however. By the 1830s and 1840s, thousands of mules were herded south through the Cumberland Gap.[67]

Farmers along the various trade routes profited from the trade, since the herds provided a ready market for their grain. Conveniently for all concerned, much of the stock traveled south in the fall or winter, after the harvest. Farmers were less busy at that time. They had their grain stocks ready for sale, and planters looked forward to spring planting with an eye on expanding their operations or replacing worn-out animals. Too, yellow fever was at a low ebb, a not inconsequential concern for Kentuckians headed toward the low-country South.[68]

Breeders sometimes traveled across a state or even to Europe to procure blooded jack stock with which to improve their mule herds, but they usually did not drive their stock to market. While breeders and stock raisers might raise a major portion of mules on their estates and reap the profits, they relied on others to transport the animals south. In Kentucky, most mules were raised by slave owners so that during much of the antebellum era, one group of slave owners in the Upper South supplied another in the Lower South with mules.[69]

Generally, mules were not bred, raised, and sold as mature animals by the same owner, but might change hands a dozen times before being driven to a southern market.[70] Individual farmers might have one or more mares bred to a jack and raise the young mules until they were weaned, but most did not have time, inclination, or capital to support excess animals to sale age. The *Southern Planter* reported that in Kentucky, Indiana, and Ohio almost every farmer kept one or two mares which were bred to jacks practically every year. Many of these mares also worked as draft animals while in foal.[71] Consequently, a degree of specialization developed in the mule business which benefited several groups, although large, specialized mule-breeding farms never developed. On the other hand, as time passed, several regional market points developed into which mules funneled.

Many farmers in mule-raising areas such as the Kentucky Bluegrass, the Nashville Basin, and later the Missouri Valley bred their mares to local jacks. Kentucky mule raisers also sent jacks into Ohio, Indiana, and the hinterlands of their own state. Farmers willingly paid the jack's fee, knowing that any produce from the union would be easily sold for a profit at six months of age.[72] At that time, graziers purchased mules from individual farmers and took them to Kentucky to raise to maturity.[73] These large stock farmers normally kept the mules on pasture supplemented with a light ration of grain. As the animals neared maturity, the feeder increased the grain ration to "bloom" his animals for market.[74]

Drovers carried out the final step in marketing the mules. Until railroads offered a better alternative, drovers purchased groups of mules from graziers to drive to market. Large stock growers might contract with drovers, men who were seldom full-time farmers but often full-time stock traders, although in some cases they were farmers who had time during the autumn and winter to take herds to market.[75] As other forms of transportation supplanted the mule drive, drovers disappeared or became mule traders. This decentralized process worked well enough to supply the South with millions of mules through the years, despite the fact that some mule buyers and dealers believed a specialized mule-raising farm could be profitable if run efficiently. Even so, as late as the 1920s, three-fourths of the mules in Missouri, which by that time was a major mule-producing state, were raised by farmers who considered mules a sideline.[76]

The practice of breeding and raising animals on different farms developed first in the cattle industry, so that mule breeders had a viable model on which to base their own version of the range-feeder trade.[77] In some cases, antebellum farmers borrowed money to purchase a group of mules to eat the farm's excess grain. It was more profitable to do that than to sell the grain, for the farmer knew mule traders would soon be about offering him a modest profit for his trouble.[78] In Mercer County, Kentucky, an observer reported what was doubtless true for much of the Bluegrass region in the 1850s: that mules were the most profitable stock raised in the area.[79] Similar arrangements developed in other mule-raising areas such as Tennessee, Missouri, and, later, Texas.

The Civil War decimated the southern mule population, stopping trade from Upper South to Lower South. After the war, jacks in the South were for the most part worthless because mule breeding throughout the region had been ruined by the hostilities, the jacks had aged several years, and mares suitable for breeding were scarce.[80] Likewise, jack breeding declined precipitously during the war and did not resume in earnest until the 1880s, when a boom in mule breeding occurred with a large number of imported animals serving as the foundation for the newly invigorated mule-breeding industry.[81] Kentucky and Tennessee breeders imported large numbers of European jacks during the late nineteenth century, but significantly so did breeders in other states such as Missouri, Illinois, and Nebraska. After approximately 1885, mule breeders imported larger numbers of better stock than ever before.[82] Such was the renewed interest in mule breeding and jacks during the 1880s that for the first time mule breeders organized a registry association for jacks.[83] Importers spoke not of purchasing individual animals, but of buying jacks and jennets in lots.

The focus of mule production slipped westward after the Civil War with Missouri and Texas rising to great prominence as mule-producing states.[84] Mule raising in Missouri dates back to the importation of animals along the Santa Fe Trail from New Mexico after 1821.[85] By the 1850s, Missouri was one of the chief mule-breeding areas in the nation, with its mules sold mainly in the South and in California and Oregon.[86] Between 1910 and 1920, Missouri mules increased in value by 185 percent and the state was recognized as the "boss mule state of

Fig. 3. Prize-winning jacks at the Missouri State Fair, 1911. (*American Jack Stock Stud Book*, vol. 9; Courtesy of the American Mammoth Jackstock Registry)

the Union." The rise of Missouri as a mule state rested upon several factors—good pastures of bluegrass and clover; relative cheapness of farmland; a climate suited to keeping animals on pasture much of the year; and the nearness of "the two best mule markets in the world," St. Louis and Kansas City.[87]

While Texas would not play a role as a mule supplier to the cotton-producing regions as early as other states, the region had a long history of stock raising of all kinds. Mules were especially important to Texas because of its proximity to Mexico. Before the Texas Revolution, mules were used to transport goods between Texas and Mexico, and with the low cost of raising them, mule breeding and selling was considered one of the most lucrative pursuits in Texas.[88] Texas mules were also important to areas such as Louisiana as early as 1800. In the years following this date, horses and mules overtook cattle as the main export from Texas to Louisiana. Even when this trade was made illegal by Spanish law after 1803 when the United States came to possess the Louisiana Territory, the movement of horses, mules, and cattle continued because of the high profits involved.[89] Thus, trade between Texas and some points in the Lower South had long been established when mule use increased in the antebellum period. As mule use increased after the Civil War, Texas logically came to play a larger role in supplying mules to the region.

Mule traders linked the mule-producing states with their markets. These traders played a crucial role in supplying the South with draft animals, and it was men like Ray Lum who possessed the experience, expertise, and savvy to determine or set a mule's worth.[90] Mule traders comprised an important segment of southern society, but they left few records since much of their business rested upon handshakes, oral contracts, and, at times, noncash transactions. Many men made their livings from stock trading, and to be southern was to be a mule trader of sorts. Kemp P. Hill spent a lifetime buying, selling, and trading animals and believed practically "every man in the South who did any kind of business at all had to know something about mules and horses."[91] Mules made up a large amount of the conversation among farmers, and being a good judge of mules carried much weight in a community.[92] Not surprisingly, a plethora of folk tales and jokes revolved around mule traders and mule trading.

The mule dealer developed as demand for mules increased and the entire mule-breeding, mule-raising, and mule-trading industry matured. During the antebellum era, large mule dealers were often Kentucky and Tennessee planters. According to one mule breeder, these men felt duty bound to deal openly and honestly with potential customers, for the sake of both business and their good names. If their animals "had blemishes that were not perceptible they would point them out to you," he noted. To do otherwise would damage the seller's reputation.[93] Of course, competition existed among mule breeders and traders during the antebellum years, such as that exemplified by Mark Cockerill's less than subtle touting of Tennessee mules, but the explosive demand for mules tended to provide a ready market for the animals that could be produced, thus removing some competitive pressures.

As mule trading increased to meet demand during the twentieth century, more competition developed among those involved. Key geographical centers such as Nashville, Tennessee, and Kansas City, Missouri, drew aggressive and ambitious mule traders such as Ferd Owen. Local mule traders like Kemp Hill might travel to the large market cities to buy mules, but their business remained tied to one area. One mule trader recalled that southern buyers from Georgia, Kentucky, and Tennessee regularly came to Warrensburg, Missouri, to purchase mules. Traders also traveled to the major markets to sell their animals. Keith Jones related that in Richmond, Virginia, in 1937 "[i]t wasn't nothin' to sell 7–800 mules a week there." He also shipped mules from Missouri to Deckard and Chattanooga, Tennessee.[94]

Fig. 4. Mule dealers such as Mississippi's Ray Lum played a central role in supplying southern farms with draft animals. (Courtesy of Mississippi Department of Archives and History)

The owners of the South's largest operations might send agents to purchase mules. The huge Mississippi Delta cotton plantation called the Delta and Pine Land Company had its own mule manager who traveled widely in search of replacement mules.[95] Even with professional care and proper handling, Delta and Pine Land Company's mules did age, get sick, and die. The plantation had to replace approximately 150 mules each year.[96] Delta and Pine Land Company's livestock manager, J. W. McCluer, traveled to mule-producing areas to inspect and purchase replacement stock. Depending on the plantation's financial situation, plantation manager Oscar Johnston would wire money when needed. On one trip to Missouri, McCluer purchased around sixty mules for $7,000, but Johnston advised him to purchase no more at that time since the cotton sales had "slowed up" and the plantation was being called on "to take care of our obligations to the bank." At this point the plantation had purchased only one-half of the replacement mules it needed.[97]

Since Johnston chose to remain dependent on mules, he sought to make mules an efficient part of the overall operation of the plantation, and Delta and Pine Land's mule-use policies illustrate much about a mule's existence once purchased by a plantation, as well as how heavily southern cotton culture depended on the mule trade. Johnston rationalized mule use on Delta and Pine Land acres by centralizing mule handling at key points. Twelve large mule barns housed about one hundred animals each. Johnston believed such organization cut feed costs and assured each tenant family a healthy, well-fed draft animal.[98]

The plantation's organization benefited both landlord and tenant. Johnston estimated that his method of controlling and distributing mules reduced the number of mules and concomitant mule equipment by twenty-five percent compared to the quantity needed if each tenant supplied his own mule and equipment. Besides saving on feed costs and depreciation of the mules, Johnston noted that the company's mule policy "inures to the benefit of the tenant in that he is not called upon to have any investment in this equipment and is paid cash for the labor put in by him in working plantation feed crops, thus supplementing the income from his cotton crop." In addition, Delta and Pine Land did not require tenants to have their own gardens or small animals, but it did encourage them to do so. A "considerable" number of tenants had their own gardens, chickens, and pigs. About half of the tenants owned milk cows as well.[99]

Because the plantation had to absorb losses from an animal's sickness or death, mule care was a key element in Delta and Pine Land's overall strategy to cut costs. Healthy animals worked better, wasted less feed, and lived longer. Minor savings multiplied and added up to substantial gains for the plantation, especially when it cost nearly $120 annually to feed each animal in the mid-1920s.[100] Johnston was serious about the care given to the plantation's mules, as well as how hard they worked. Replying to an invitation to have some of the plantation's mules in a Rosedale, Mississippi, race, Johnston explained, "In the first place I doubt if we have any 'racing' mules; in the second place all of our mules are pretty busy at the moment 'racing' up and down cotton rows trying to make this year's crop."[101] He issued specific orders to protect Delta and Pine Land's investment. For example, the plantation had ninety-three unbroken mules which were approximately one year younger than normal for beginning earnest labor. Johnston reminded his managers to use particular care when handling this stock. Johnston's memo cautioned managers not to rope, corner, or chase the mules to catch them, but to employ a chute to catch, handle, and gentle the animals. After a period of gentling, the mules could be used for light work such as manure hauling or stalk cutting. Johnston was adamant that the mules not be used in four-mule teams for heavy work such as plowing until they were mature enough to handle such jobs.

Early in the handling and training process, the mules received a name. No doubt many of the names were assigned arbitrarily, but some mule names perhaps reflected a dominant character trait. Among Delta and Pine Land's work stock inventories were Cannonball, Solomon, Dude, Sudden, Shock, Doll, and Meningitis. As the groups of mules matured, Johnston removed the limits on what the animals could be asked to do. Throughout their tenure on Delta and Pine Land acres, mules were carefully tended by the plantation veterinarian. Older mules incapable of work were "deadheaded" and sold, but those capable of some work were categorized as "half mules."[102] In 1940, some ninety mules fit the "half mule" category.[103] Johnston not only desired the best for the young mules but also expected each mule to work as long and as hard as possible.

Johnston took steps to help curtail high mule-replacement costs. First of all, he began purchasing young rather than mature mules in order to "reduce substantially the annual cost of mule replacement." During the summer of 1939, the plantation purchased 150 mules of approximately fourteen months of age at an average cost of $123 per head. These mules began work in 1941. Another 150 mules only seven months old were also purchased during the same summer for $78 each. The mules in this lot were not destined to work until 1942. In mid-1941, Delta and Pine Land Company bought another group of 150 seven-month-old mules for less than $62 per head for the 1943 work year. The success—that is, the monetary savings—of the program pleased Johnston. The five- or six-year-old mules normally purchased averaged $230. By feeding young mules until mature, the plantation realized substantial savings. Cheap feed was the key to the program. The mules foraged on plantation pasture seven or eight months of the year, but needed to be fed grain and hay the remainder of the year. Delta and Pine Land's plantation possessed significant acreages withdrawn from cotton production as well as land deemed unsuitable for cotton production. Significantly, some of the acreages that provided mule feed had been taken out of production under the New Deal cotton-reduction program.

The use of lands taken out of production provides an intriguing paradox. Instead of fueling tractorization on Delta and Pine Land Company's lands, cotton-reduction programs entrenched mule use during the Depression decade. This happened despite the fact that the plantation received huge government payments for reducing its cotton crop. During the first three years of the Agricultural Adjustment Act (AAA), Delta and Pine Land received more than $318,000 from the federal government.[104] On otherwise nonproductive lands, "considerable quantities of rough hay" could be grown with which to feed the larger than normal number of mules at minimal cost. The hay could not be sold for two reasons. First, most farmers in the area produced their own hay and shipping costs precluded any profit from sales outside the immediate area. Second, crops grown on acres taken out of production under the auspices of the AAA could not be sold. Thus, the plantation created a market for its own hay.

By 1941 Johnston believed the mule purchase program was beyond the "experimental stage" and reported that it was "definitely profitable."[105] In the case of Delta and Pine Land Company, substantial AAA payments did not go toward purchasing tractors, so the New Deal may have slowed mechanization on this plantation, although Department of Agriculture programs catalyzed mechanization in other cases.[106] Oscar Johnston had begun experimenting with tractors before passage of the AAA, and only in the mid-1940s did the mule population begin to decline as mechanical equipment replaced draft animals just as Johnston's program to reduce mule-replacement costs was proving itself.

Mule trading often required much travel, both to distant markets and in the dealer's vicinity. Kemp Hill moved constantly in the 1880s, buying, selling, and trading animals. Hill started out in the mule business at the age of fourteen with a two-horse wagon. Plodding from county to county and often camping when night fell, his stock following behind or tied to the wagon, Hill timed his travels so that he would be at a county seat when court was in session. Practically every farmer in the area made it to town sometime during the week, providing Hill with ample opportunity to ply his trade. This, Hill recalled, "[m]ade a pretty nice arrangement for us, and we did plenty of trading."[107]

Traders might move widely within a state or several counties, but many also traveled to the mule-producing states to obtain animals. Some planters such as Joshua Lee's grandfather acted as de facto traders for their areas. When "Cap'n Tolbert" went to large mule sales to buy mules for his own fifty-plow operation, other farmers saddled him with as many as one hundred orders for mules.[108] Such men differed from full-time mule traders, however, since they bought mules at the large auctions but did not travel to mule-producing areas to purchase mules. Doubtless some planters did purchase mules directly from Kentucky, Tennessee, Missouri, and later Texas, but most relied upon the large markets such as Richmond, Chattanooga, Memphis, and Atlanta which were supplied by mule traders who specialized in gathering large herds for shipment to the trading centers.

Local mule traders relied upon the large suppliers as well, especially after railroads linked southern towns with the region's cities and other parts of the country. Kemp Hill, who claimed to have owned the largest stables in his area of North Carolina, took the train regularly to St. Louis, Missouri, to purchase a carload of mules. Hill, like most traders, loved to deal and did not hold his stock until he arrived home. On the return trip, towns like Atlanta, Georgia, and Greenwood, South Carolina, served as rest stops for the stock. Hill made good use of his time at layovers, selling and buying mules as he saw fit, sometimes for several days at a time. More than once he sold all of his mules and had to return for another carload of animals to take home.[109]

Buying twenty-five, thirty, or more mules required large amounts of cash, but shrewd traders could make handsome profits. Early in his career as a mule trader, which began in the 1880s, Kemp Hill was, like the South, cash poor. Consequently, barter became a standard of trade. For his part, Hill began by trading mules for scrub cattle which he sold to Virginia and Maryland farmers for cash. Upper South stockmen fattened the calves for sale. He also took timber "or anything that we could turn into a profit." "All I wanted," explained Hill, "was the cash so I could go on out to the western markets and buy up more mules." Eventually Hill and the South acquired enough capital so that mule trading shifted to "simpler and cleaner" cash transactions. Later in life, Hill diversified his interests by getting involved in cotton futures, farmland, and timber. Some years he made as much as $25,000, but those were balanced with stretches of up to two years when he cleared no profit.[110]

Trading of any type had its allure. Hill invested in volatile futures and land markets, and he noted one could make "pretty money when you hit it right." Mules, however, usually offered solid returns. During the early twentieth century, Hill paid around $50 per head for mules. He paid the railroad approximately $50 to ship the stock to North Carolina and then sold the animals for $175 to $225.[111] Mule demand fluctuated, however. The early 1930s were especially tough for mule traders. Hill went broke in 1930 when mules simply would not sell.

Scores of small mule dealers worked in the South, most of them handling mule sales for a limited area, but a few rose to national prominence. If Kemp Hill represented the small, local mule supplier, Ferd Owen defined the large mule dealer. Ferdinand Lincoln Owen dominated the mule trade in the 1940s as did no other individual. Owen began trading mules and horses early in his teens, having learned how to assess the merits of the animals from his father. By the age of fifteen Owen had begun working on his own as a "road trader" or "gypsy." Like Kemp Hill, Owen traveled in a wagon with his mules and horses following behind. Based in Arkansas, he worked north into Iowa and south into Oklahoma. Eventually he oversaw the Owen Brothers Horse and Mule Company, the Ferd Owen Horse and Mule Company, and the Kansas City Horse and Mule Company. Owen's companies sold mules nationally and internationally. In 1946 alone, Owen bought and sold approximately one hundred thousand animals. The ability to buy and sell such quantities made Owen a potent force in the industry. One author concluded that Owen "starts the market, protects the market and generally speaking, is the market." On heavy trading days at his huge Kansas City, Missouri, barn, Owen might purchase fifteen hundred animals, thus setting prices for the nation's mule trade. He estimated that over his forty years as a trader, he had handled a half-million animals.[112]

Much of Owen's success grew out of his years of experience, his memory, and his talent for trading. An astute trader, Owen enjoyed both the give-and-take of one-on-one negotiating and fast-paced auction sales. He seldom lost money on a mule, but knew what it felt like to be cheated. At twelve years of age, Owen bought a mule that was blind in one eye and losing sight in the other. Owen's father refused to help him get his money back, counseling him instead: "If you get skinned, learn to sit on the blister." Owen took this to heart, as did most mule traders, and often did the skinning as he gained experience. Mule trading was no holds barred. "I don't mind robbin' somebody if they come in to rob me," explained Owen after one sale.[113]

Owen also relied heavily on his family to help run his business. His wife played a key role in building up the business by keeping the books, teaching Owen to read and write, and even helping to break green horses. Eventually, seven of Owen's nine brothers, as well as a host of cousins, in-laws, and other relations, participated in the family's mule business. During the mid-1940s, the Owen clan handled about fifty percent of all public horse and mule sales in the nation. Owen invested his earnings wisely during his career. For example, in 1945 he purchased a 2,038-acre Cass County, Missouri, cattle ranch as an investment. He lived, however, in Clay County, in a twelve-room, handsomely furnished mansion. Owen excelled in an extremely competitive calling.[114]

By the twentieth century, mule traders had replaced the grazier and drovers of earlier times. Better land and river transportation limited the need for "gypsies" to travel through mule country with their herds in tow. Although railroad shipping was faster than driving the animals to market on foot, it demanded certain precautions of its own. One trader related that when shipping mules by rail, he made sure they had no shoes on their rear feet in order to limit the damage they could do to one another when kicking. He also put coal oil on their tails to keep them from chewing and bobbing one another's tails, thus driving their prices down.[115] As hard-surface roads linked farms to markets, farmers who lived a reasonable distance from central markets such as Nashville or Columbia, Tennessee, bought trucks and moved livestock to and from market themselves. Major mule markets such as Nashville, Tennessee, played a growing role in the mule trade.[116] Conversely, farmers could buy mules and return to their farms with their animals loaded in the back.

Mule trading became more centralized, more formal, and less itinerant as the twentieth century progressed. Certainly old forms died hard, such as the court-day sales and the drovers moving south with large herds, but the transportation developments dictated some changes in the way mules moved from the mule-producing states to the Lower South. Railroads certainly played a large role in the rise of Missouri and Texas as major mule producers since the rails made moving large numbers of animals long distances logistically and economically feasible. Farmers who needed only a mule or two could band together and purchase a railroad car of mules, saving money in the process. Extension agents could help organize a county's farmers in this way, but as likely as not the local mule dealer played a role, too. County agents seldom tried to circumvent traditional avenues of purchasing mules. No doubt they preferred to work with farmers and dealers, rather than putting themselves in the middle of a potentially difficult situation by acting as a buyer or seller.

Mule auctions arose as large numbers of farmers, traders, and dealers converged on central shipping points. Whereas mule traders had once dealt with many farmers personally, the large number of mules southerners demanded led to a less personal type of mule trading, at least at the large markets. On the local level, farmers might "dicker" with a mule trader one on one, but the large sales at barns like those at Nashville or St. Louis precluded much of a personal approach. The animals became the focus at auctions, not the individual buyers and sellers.

Auctions are still a staple in many southern communities, where one can witness the rapid sale of numerous animals on the night of the "sale." Although slave auctions have deep roots in the southern past, it is unclear when horse and mule auctions became commonplace. In some areas, auctions developed rather late in the mule era. Nashville, Tennessee's animal exchange did not implement a horse and mule auction until 1938, for example. Horse and mule auctions at major exchanges were characterized by rapidity and, if necessary, length. At Nashville's horse and mule auction, bidding began at ten o'clock in the morning and continued well into the night during the heaviest trade days. More than 50 buyers might be on the sale floor, while 75 to 150 spectators watched in the stands as individual animals or teams moved through the sale ring at the rate of 75 to 100 every hour to the staccato bark of the auctioneer.[117]

The local mule barn, such as the one described by Georgian Harry Crews, was an entrancing place for a young boy. Mule barns were male-dominated worlds. Apart from the mules, the men discussing farming and mules, and the general feel of excitement, Crews remembered that there "were never any women at the mule barn. This was the place of fathers and brothers and uncles, a quintessentially male world, and for that very reason a place that was almost unbearably pleasant for a young boy who, although he did not know it, was learning the ways of manhood."[118]

Ferd Owen acted as the chief auctioneer at his sale barn in St. Louis. Like any auctioneer, Owen had to determine the worth of the animal in fifteen to twenty seconds. A quick look told him the size, weight, and approximate age of the animal; whether it was sound or not; its disposition; its blemishes; its probable utility; its price; and his potential profit if he should buy the animal. Owen normally rapped each animal on the hindquarters with his cane as it left the ring, unless he happened to be eating peanuts, in which case he would throw a handful of shells at the animal. He also kept up a lively banter with the buyers.[119]

Not all buyers bought at auction, however, since they believed that "the only function of an auctioneer was to talk dollars onto a mule's price tag." Those averse to the sale ring might arrange to purchase mules among the stalls, where negotiations could be carried on more quietly and where the merits of an animal might be judged more closely. Joshua Lee's grandfather, for example, liked to go into the stables at auctions rather than compete in the auction barn. He believed he could better judge a potential purchase by entering the mule's stall and flipping his hat at the animal. He would not buy a mule that cringed or shied, but would further examine one that reacted little to determine if it was hardheaded, lazy, or simply calm-natured.[120]

Ferd Owen became more successful than his fellow mule traders, but in many ways he exemplified the type. Mule traders doubtless exuded individual quirks as varied as the animals from which they profited, but many had common traits. Perhaps first and foremost, mule traders loved to buy and sell, to "deal," to "trade." Owen even sold some of his own work mules in the field, in harness, as they harvested grain. As he explained, "'I'm just a natural-born trader man—never knew nothing else—never did nothing else—never had nothing else on my mind.'"[121] Trading mules had its own attractions as well. Mules sold readily at any age, although dealers preferred unbroken mules. Thus, mules did not have to be carefully groomed before sale or broken to saddle or harness. It was simple to get mules groomed for sale. Their manes were clipped close, from about three-fourths of an inch at the middle of the neck tapering to one-half inch at the withers and ears. This was sometimes labeled a "roached" mane. Mule tails usually got clipped of all hair from four or five inches down from the base of the tail.[122] Mules also sold well in large lots, and dealers could get a bulk payment which they could invest to better advantage.[123]

Obviously a trader's success rested upon whether he could tell a good animal from a bad animal. With mules, buyers often had to rely upon determinants other than watching the animal work since mules generally were not broken before being sold for the first time. This practice dated back to the antebellum era when the *Southern Planter* noted that planters "are too cautious to buy a broke mule, lest it should prove to be an antiquated, broken down beast, fattened up, and sold for a young one,—as it is more difficult to judge of their ages than that of a horse. The external marks of time, and service is not generally so apparent upon them [as horses]."[124] What was true in 1858 was true in the twentieth century when mules were bought and sold in large lots. Caveat emptor applied universally in mule dealing, for indeed some unscrupulous traders bought old mules, fattened them up, and sold them as young animals.[125]

Dealers also had to have a keen appreciation for the animals that might have been neglected. It was relatively easy, in other words, to select a good mule when it was in good condition, but it took "a smart man to tell a good mule when he is thin." Some farmers wasted little grain on their mules, since they raised them for sale, not for work. The buyers were expected to fatten the animals out of their own pockets. Thus, they had to have a practiced eye to select the best animals to fatten for sale. Mule dealing required, therefore, that its practitioners be shrewd. Hard traders, they nonetheless came to one another's aid during hard times. Finally, mule dealers tended to be stylish in their dress and particular about their grooming habits.[126] Ferd Owen favored dapper suits, a hickory cane, and a diamond stickpin shaped like a mule's head.[127] Mule trading was both art and science, a calling in which the dealer had to judge both man and animal.

Mule trading could also require substantial capital. The aggregate cost of travel, buying numerous animals, keeping them fed and watered, grooming them, and shipping them to market was daunting. The yard in Nashville, Tennessee, charged for its services, as did any auction barn. Selling charges in 1946 were $2.50 for each animal sold or $1 if the animal did not sell, five cents for insurance, thirty-five cents for yardage, and sixty cents per day for feed. The yard also provided halters for forty cents and would trim a mule's mane and tail for twenty cents. Trims for horses cost but a dime.[128] For a farmer with only a mule or two to sell, these costs cut his profit slightly; but multiplied by a hundredfold or two, the costs involved in mule dealing could be significant. Kemp Hill always tried to sell a carload of mules as quickly as possible because feed costs "could turn out to eat up a lot of profit."[129]

In Ferd Owen's case, these costs were staggering because he handled such large numbers of mules. During the 1940s, for example, he handled between one hundred thousand and two hundred thousand head of horses and mules each year. At one point, Owen lost $250,000 on a major mule sale because red tape forced him to hold several thousand mules destined for Mexico. Since feed, hay, and water costs amounted to fifty to sixty cents per day, Owen estimated that the mules ate up all his potential profits on the deal in about two weeks. Eventually he closed the sale of twelve thousand mules to Mexican purchasers for more than $1 million, but he claimed to have profited little, if any, from the sale. Owen's competitors attempted to have a bill pushed through Congress forcing sales of the type Owen negotiated to be approved by Congress. The bill failed, and Owen sold his mules to Mexico, but not before he had been forced to reduce the number of mules sold and the price he received for them.[130]

In spite of the great success of mule traders and dealers, however, even a dominant position in the mule market could not save traders as farmers shifted away from mules and began buying tractors. Farmers bypassed auction barns and stockyards and began negotiating with tractor dealers for their draft power, sometimes trading in their mules for tractors. By the middle of the twentieth century, most southern farmers had begun thinking about draft power not in terms of teams and hitches, conformation, and heart, but in terms of manufacturer, horsepower, and the bottom line. Before tractors replaced mules on southern farms, however, there was a significant effort to make certain the region would be able to maintain an adequate supply of draft animals by raising them in the cotton South.

3
An Unrealized Dream
Local Mule Production

As more southern planters switched to mules during the nineteenth century, the mule supply became increasingly important to the region's economic livelihood. During the late antebellum era, mule production, sale, and purchase became a significant topic of discussion in the agricultural press. In the two decades immediately before the Civil War, southerners discussed widely the need for an agriculture less focused upon staple production. Farmers spent their lives growing cotton, "merely to buy everything else," observed the *Southern Cultivator.*[1] Mules were an important part of the "everything else," for even though the South produced a substantial quantity of livestock throughout the antebellum period, mules were never produced on a large scale in cotton country.[2]

The weaknesses of southern staple agriculture were not lost on reformers who perennially called for a more broadly based and self-sufficient agriculture. The *Southern Cultivator* reprinted a *Memphis Eagle and Enquirer* column which summarized the reform urge by asking, "When will the South be true to herself? When will Southern men learn their true policy? When will they learn that everything produced at home—corn and horses, peas and hogs, potatoes and mules, shucks and sheep, brogans and beef—is a savings of always double, and often quadruple, of what the same article would cost, if they have it to buy?"[3]

Stock raising played an integral role in reform programs, and home mule production elicited much comment in the agricultural press. Self-sufficiency advocates were particularly intent upon convincing Deep South planters, who were most likely to have the necessary land and capital, to produce their own mules.[4] Reformers cited two related arguments for southerners to begin this endeavor. Cost was a major concern. By breeding their own mules, southerners could play a crucial role in curbing price increases. One calculation determined that if Mississippians raised one mule for every forty slaves in the state, mule prices would fall more than ten dollars per head. The *Southern Cultivator* complained and reminded readers that mule dealers already charged prices "beyond a fair value" for their animals.[5]

D. C. Graham of Franklin County, Mississippi, believed that his state's farmers "might raise many things we do not—and which we will not, as long as cotton brings its present price," including sheep, horses, and mules. Farmers raised so few horses that Graham allowed they might be counted "as none."[6] Hill Carter, addressing the Virginia Central Agricultural Society in 1859, linked mule raising and slave raising. He asked, "Why should we not raise our mules as well as our negroes?"[7]

Reformers insisted that it was much cheaper to produce a mule than to purchase a comparable animal, especially since the lineage and history of the homegrown mule was certain. Mules were said to be as cheap to raise as cattle, and as little trouble.[8] One estimate placed the cost of raising a mule to the age of three years—that is, old enough to be "put in the plow"—at $50.[9] Another assessment found the cost of raising a mule to three years of age to be $30.[10] A thrifty-minded Floridian believed that the cost to raise a mule to the age of three was approximately $35, at which time the animal was worth between $50 and $75.[11] "I defy any man," wrote a self-styled expert on mule raising, "to prove . . . that they cannot raise a better mule for $50 than they can buy at $75."[12]

A related benefit, so went the argument, which sprang from a rising number of home-produced mules, was the proportionate reduction of the South's reliance on outside sources. But this was possible only if the South increased the number and quality of its pastures, and this issue was a considerable stumbling block to the larger question of agricultural diversification. In some ways, southern diversification had to wait on a convergence of governmental, scientific, and technological developments, many of which did not occur until the second half of the twentieth century.[13]

Pointing out an obvious weakness in the southern stock-raising program, the *Southern Cultivator* declared "our meadows and pastures should be on our own plantations; and not a thousand miles off in distant States."[14] Letters and columns touted the suitability of southern bottomlands, including acreage along "all the streams, large and small" of the region, for stock raising.[15] Not only would increased stock production help the South by partly supplying its draft animals, but the shift to stock raising would increase the value of southern farms and promote a rise in agricultural incomes.[16] A natural by-product of diversification would be a reduction in cotton acreage and a rise in cotton prices, as previously tilled lands were shifted to pasture.[17] To stimulate interest in local mule raising, advocates argued that agricultural fair premiums "large enough to be worthy of attention" be given to owners of jacks by the southern states, and that a "tariff of protection" be erected against Upper South imports.[18]

Some southern farmers responded to arguments for mule raising and began breeding mares to their own jacks or purchased one's services. A Chunnenuggee, Alabama, planter found mule raising "profitable." He put his mules to work at two years of age, each spring breaking a new team that had been raised on his plantation. He was "inclined to think that the planters would find it to their interest to pay more attention to this branch of farm economy."[19] In Edwards, Mississippi, another planter expressed much the same sentiment, for he considered "the growing of these animals to be more profitable than making cotton to buy them with." Echoing the typical disregard for the care of work stock, he reported that his costs and trouble to raise mules to working age were negligible. "My colts," he explained, "are reared in the woods, pasture and field, scarcely ever fed, and at an expense so small that I am not able to say."[20] Colonel J. M. Williams of Society Hill, South Carolina, whose 4,200-acre plantation lay along the Pee Dee River, owned more than 250 slaves. Unlike the vast majority of planters, Williams listed a jack among his livestock holdings. He also owned sixty mules and mares, the latter as necessary a part of mule breeding as the jack.[21]

While Williams may have been atypical, enough planters had begun to raise mules that the *Southern Planter* declared in 1850 that there had been "a great increase" in the number of farmers raising mules.[22] Still, most planters, even those who prided themselves on making their plantations as self-sufficient as possible, did not raise mules. Hugh Davis, who owned the Beaver Bend plantation in Perry County, Alabama, strove to have his plantation produce many of its own needs, but apparently purchased his mules.[23] As cotton prices rose steadily in the late 1830s, many planters used available capital to purchase land or slaves.[24] After the panic of 1837, few would have had the capital to spend on expensive jacks, and while local mule production saw some limited successes in the Deep South during the difficult 1840s, any progress evaporated in the heat of the next decade's cotton fever.

The cotton boom of the 1850s, when cotton prices remained high year after year, curbed much of the interest in diversification, although agricultural reformers continued to press the issue. In Alabama, as in the cotton South as a whole, some planters experimented with diversified farming, but such plans found few adherents and cotton cultivation retained its predominant place in southern fields.[25] Much of the cotton South could have reported, as did Cabarras County, North Carolina, that "[l]ittle attention is paid here to the raising of stock of any kind. Horses and mules are brought in by droves."[26] As cotton prices climbed, mule prices remained more or less constant, thus removing a crucial incentive to local mule production, even though few southerners considered mules cheap at the prices they commanded.

The issue hinged upon shifting priorities, potential economic returns, convenience, and a heady faith in cotton. When James Henry Hammond dared the world in 1858 to make war on cotton, he both struck a chord with many southerners and reflected the confidence they felt in cotton's ability to supply the region's needs. Land could be used to raise animals or grow crops. The river bottoms so suitable for pastures, and often among the most fertile lands, could be used for cotton production as well. Staple production tended to wax and wane in concert with rising and falling cotton prices. Likewise, but conversely, interest in local mule production increased in times of low cotton prices and fell off in years of high cotton prices.

Many factors worked against local production of mules, but the lack of pastures played a key role and was one of several issues limiting diversification of southern agriculture in general. Developing quality pastures was a long-term investment, while staple-crop profits were more immediately visible than the benefits of diversification.[27] Thus, staple-crop production often took precedence over other crops in boom times, and over mule production as well. Mule production required large amounts of pasture, as well as a jack, an animal that American farmers used for nothing apart from breeding.[28] Mares, of course, could be worked in the fields even while carrying a foal, but planters had to confront the possibility that draft animals might be mistreated, which was one reason planters shifted to mules in the first place. Gestating mares would be particularly vulnerable to harsh treatment and often "slipped" their foals or had miscarriages. George Washington believed many of his mares did not carry their mules full term because they were ridden at night by slaves or worked too heavily in the fields. "Night riders, and treading Wheat," he complained, "will forever deprive me of Foals."[29]

Other factors limited Deep South mule breeding, not the least of which was the fact that mule breeding was a curious business—more an art than a science. Breeders and traders from George Washington on discovered that jacks and their offspring possessed quirks that demanded particular care, and many planters and farmers remained convinced that mule production was too expensive, specialized, and troublesome to carry out on their own. For breeders, of course, jacks constituted the largest single investment, and it was with this animal they were most concerned. Jacks were legendary for their cool temperaments, and proper handling was crucial. A stallion could be ruined because of his nervousness, but a jack could be ruined as easily by his apathetic disposition and the resulting unwillingness to breed a mare.[30]

Washington discovered to his chagrin that his jack from the Spanish monarch was handsome but quite indifferent to the mares presented to him. Alluding to his newly acquired jack, Washington wrote that his "late royal master cannot be less moved by female allurements than he is; or when prompted, can proceed with more deliberation and majestic solemnity to the work of procreation."[31] Washington later discovered that by teasing a jack with a jenny, a jack's handlers could encourage his amorous advances.[32] Perhaps the jack had grown up with females of his own kind, but Washington found it "necessary to have a Jennet or two always at hand during the season, by way of stimulus, when he is in those slothful humours."[33] As knowledge of jacks increased, it became common to raise young jacks in the company of gentle mares or fillies. Jacks isolated from their own kind were more willing to cover mares when they reached the appropriate age than jacks who had been exposed to the wiles of a jenny.[34] Jacks, in short, preferred their own kind as if they knew that the result of a union with a mare would extend their own line no further.

Jacks had their own peculiar faults, and breeders attempted to offset them by carefully selecting brood mares.[35] For example, breeders early recognized the phlegmatic nature of jacks and attempted to introduce warmer blood through the dam to produce a more active or spirited offspring. "To produce fine mules," recommended the *Southern Cultivator,* "too much attention cannot be bestowed in selecting the mares, which should be the best stock that can be obtained—combining good blood, size and fine symmetry."[36] In short, antebellum breeders understood that quality jacks bred to quality mares would produce the best mules and that they needed to use the same care in selecting breeding stock for mules as they did when choosing stock to produce race or trotting horses.[37]

The mule's chief champions, exemplified by men like George Washington and Henry Clay, differed from the mass of breeders in their attitudes toward the art and science of mule breeding. Washington and Clay looked at mule breeding with an eye toward profit, but they also were intent upon producing the best possible mules by using the best available breeding stock. To them, profit and quality went hand in hand. However, many more mule producers, especially as the demand for mules soared in the Deep South, cared little to none about the quality of their product. Mule production to them was but a sideline, and one to be pursued at the least possible cost and with the least possible trouble. As a rule, any jack would suffice to breed any mare. Mules would sell, and the mule breeder would not be personally identified with a particular animal because of the distance between producer and buyer.

Despite some writers' hyperbolic descriptions of jacks dominating the breeding of blooded mares, mule breeders found it much easier to talk about uniting high-quality jacks and mares than to carry it out consistently since a large number of mules, perhaps most mules in the antebellum South, were products of jacks and "common coarse mares."[38] This trend continued throughout the mule's tenure as the South's premier work animal. Breeders took particular pains to select mares for producing workhorses, but often exercised much less care when breeding to produce mules. A general rule allowed that mares unsuited for breeding to stallions were good enough to produce mules.[39] This proposition died hard despite recommendations that the "nervous, rattle-headed, long-backed, slab-sided, ewe-necked, pigeon-toed, brittle-footed, blemished mare has no mission save to be worn out and forgotten." This author goes on to assert that two good horses would work longer and more cheaply than three poor horses.[40]

Even in the mule's heyday, the animal received little respect from many breeders who viewed mules as a suitable, inferior by-product of horse-breeding programs. A leading draft-animal advocate suggested that to improve large draft-mare production, small draft mares "or those that are particularly unsound or defective in conformation, be bred to jacks, for all horsemen are in accord that such defective mares should be eliminated from perpetuating their kind, and if they are diverted to the production of mules their defects will die with them."[41] Most of the "defective" mules, of course, would be sold south to plod among the cotton rows, and then be held up as examples of the inferiority of the region's agriculture. In an attempt to improve the foundation stock, mule breeders founded the Standard Jack and Jennet Registry in 1908. Until June 1, 1920, jacks and jennets could be registered with the organization regardless of ancestry, if they met size requirements—14½ hands in height with a 63-inch heart girth for jacks; 14 hands in height and a 61-inch heart girth for jennets. After June 1, animals not enrolled were barred from the registry. The Registry reminded breeders that buyers demanded registered stock.[42]

Jacks were generally smaller than the mares they bred, and this could lead to obvious logistical problems. Mares too large for the jack to mount were commonly backed into a shallow depression. The union could have comical overtones, especially if the mare was especially well bred and the jack small and common. Even so, some jacks had an entrancing way with mares. Harry Crews recalled one donkey named Jack which possessed a particularly impressive aura.

"No king," Crews noted, "ever approached his queen's couch with greater dignity or delicacy than Jack approached a mare." As the mare was readied to receive her consort, "Jack would approach in a little prance that was at once finely measured—almost mincing—and utterly confident. He would raise his great head and, with his upper lip extended, breathing like a bellows, he would test and taste the air about the mare. And she, nervous, a little beside herself, would turn to look at him over her withers with nostrils flaring, flanks trembling, ready to bolt. But he would soothe her, calm her, and caress her with the gentlest little nips along her ribs with his great, strong teeth."[43] While most encounters were probably not so poetic, so went the union that would produce a mule.

Although some breeders could wax poetic at the encounter of a jack and mare, the trouble of producing mules outweighed the benefits to many southerners in the cotton-producing states. Numerous considerations, then, tended to work against local mule production, especially in the Deep South. Breaking the Lower South's dependence on mule-producing states would have taken more than appeals to reason. Even during the antebellum era, it was clear that suitable draft horses and mules could be raised in the South, "but a large majority" of planters preferred to buy draft animals from outside the cotton South.[44] Southerners demanded more mules throughout the late antebellum period, and in response a constantly growing supply of replacements streamed southward from mule-producing regions of the border South.[45] After the smoke of the Civil War had cleared somewhat, mules once again became a major item of trade from regions outside the cotton South. Besides Kentucky and Tennessee, Missouri took its place as a major mule-breeding state.

After the Civil War, debate in the agricultural journals over the merits of mules, horses, and oxen ceased. Other issues took precedence, ranging from the state of the southern economy to the advantages and shortcomings of the new landowner-laborer relationship. Yet, some southerners continued to call for home production of mules after the Civil War. One writer to the *Southern Planter* hoped that the insecurity of the postbellum years would finally bring southerners to their senses and create an environment in which they would produce more of their own mules "by a little trouble and timely attention." "One or two [mules] raised upon a farm every year would cost comparatively little," he concluded, "and save very much in the end."[46] The *Southern Cultivator* argued in the same vein the next year. The author asserted, "The truth is . . . there is no difficulty in raising mules at the South, and very cheaply too." But the writer touched upon a crucial point by adding that except for the costs of the mare and the all-important service of a jack, "the cost of rearing a mule is very little greater than that of raising a cow."[47]

Throughout the late nineteenth century and twentieth century, the cotton South continued to rely on outside suppliers to meet the demand for mules, although some halting steps were taken to raise mules in the Deep South after 1920. National mule production dropped significantly enough during the early 1920s that observers predicted a serious shortage of work stock on farms that depended on mules. World War I absorbed some animals, food production took precedence over breeding, and the movement of midwestern farmers toward tractorization all affected the mule supply, since many of the mares bred to jacks lived in the Midwest.[48]

Draft-animal production could lag several years behind demand because of the time it took to raise animals to working age. According to one author, the mule population in the South in the early 1920s was "just beginning to recover" from the impact of World War I which took so many animals for the Allied war effort. As a result of the South and West being "scoured" for mules and horses from 1914 to 1916, draft animals were scarce and costly.[49] Farmers had a difficult time deciding when to increase or curtail production, and hitting the market at the right time was at best an uncertain proposition. Although mule breeders could usually expect to receive higher prices for mules than for horses, price changes still affected production of the animals. But it was the faltering supply of mules that pushed some southerners to consider raising their own mules.

Farm journals such as the *Southern Agriculturist* and the *Progressive Farmer* encouraged farmers to breed mules because the shortage conditions clearly pointed to a rise in prices in the near future. The *Southern Agriculturist,* for example, reminded farmers that "the demand for good mules is better than the demand for any other workstock" because mule prices stayed higher and steadier than draft-horse prices throughout periods of low and high supply.[50] In a similar vein, *Progressive Farmer* warned of a coming horse and mule shortage in the South, which was to be exacerbated by an aging mule population in the region, and it called on Corn Belt farmers to raise more draft animals for sale in the East and Southeast.[51]

Ferd Owen, the nation's dominant mule dealer of the 1940s, related that most of his business until the 1920s revolved around working horses. The rapid shift of midwestern farmers from draft animals to tractors during that decade destroyed much of the midwestern horse market, however, and Owen turned to mules, which remained in high demand because the South had not shifted to tractors on a significant scale. By 1945, Owen was paying $250 to $275 for a good-quality mule, but only $40 to $50 for a prime workhorse.[52]

Agricultural leaders for nearly a century had tried to convince southern farmers to breed mules using only the best stock, but had achieved few results.

During the twentieth century, the United States Department of Agriculture attempted to step in and succeed where agricultural reformers and journalists had failed. Under the guidance of Seaman Knapp, the USDA set up a network of county agents to help educate and guide farmers in improving their agricultural practices. In essence, during the early twentieth century the extension service, through cooperative agreements among local, state, and federal agencies, established a system of practical education for farm families in agriculture and home economics.[53]

The USDA ambitiously sought to make the cotton South less dependent on outside mule suppliers, even though mule breeding was never its primary focus. The task appeared daunting, to say the least. Many county agents reported little or no interest by farmers in participating in local work stock–breeding programs. One Georgia agent noted "[p]ractically no interest in breeding." Another agent admitted his inability to generate interest in mule raising, while yet another believed his area's farmers to be "way behind in raising horses and mules." Mule raising apparently held as little interest for black farmers as it did for white farmers. For example, the agent assigned to work with blacks in Amite County, Mississippi, wrote that work-stock production was "of little interest" in his county.[54]

Home production of mules seemed a logical answer to many of the farmers' complaints, but southerners had chosen to resist any moves in that direction. Herein lies an ironic twist to southern mule breeding. One of the most interesting features of the extension system's role in the mule-production effort was that the USDA usually prided itself on its modern vision. But in the case of mule breeding, the county agent—normally acting and seen as an agent of modernization, innovation, and change—spearheaded the movement for home mule production. While mules were not viewed as innovative, extension agents supported the contention that southern work animals needed improvement before certain shortcomings of southern farming practices could be addressed. Extension workers reported the inability of the generally "low-bred," lightweight type of mule so long used in southern fields to perform good field work. Since little difference existed between the cost of raising and working low- and high-quality mules, it made sound economic sense to push high-quality mule production on the local level.[55] Thus, the USDA at once attempted to alter a fundamental aspect of southern agriculture by encouraging dependence on a traditional draft animal.

During the early years of extension work, county agents faced considerable difficulty in gaining farmer confidence, even as the boll weevil decimated larger cotton acreages each year in its trek eastward. Once established, boll weevil eradication programs absorbed the bulk of extension energy.[56] Mule raising, then, received little attention from extension agents before the 1920s, although at the national level there had been a recommendation as early as 1917 to encourage mule breeding in the Cotton Belt.[57] The success of importation from mule-breeding states also tended to limit farmers' interest in raising their own work stock. A perennial dearth of good pastures limited what the majority of farmers could do toward raising any type of stock. Few, therefore, opted to experiment with unproven breeding programs. Southern farmers, although familiar with mules as draft animals, knew much less about raising them successfully.

Still, a number of Deep South farmers did breed mules. In some areas such as Campbell County, Georgia, a handful of farmers raised some colts from mares they used as work stock without any prodding or guidance from the county's extension agent. Since farmers raised so few draft animals, they usually found an adequate number of sires already scattered throughout the region, at least until the 1930s.[58] And while Deep South farmers did not as a rule raise their own work stock, when they did, they preferred mules for two reasons. Farmers believed mules were easier to raise than horses, and mules usually sold for higher prices than horses. One Georgia farmer estimated that he could raise a mule as cheaply as he could a cow.[59] This rule of thumb applied to South Carolina as well, where some farmers told their extension agent that they could raise a mule to two years of age as cheaply as they could a steer.[60]

County agents argued that the key to improving southern work stock was to produce it locally, where quality could be developed, maintained, and ensured through careful breeding programs. Despite evidence that some farmers raised mules, however, most southern farmers did not raise their own work animals. The number of jacks in the South remained low well into the twentieth century. In 1920, for example, Georgia farmers claimed but 427 asses and burros on their farms. Other southern states reported the same lack of breeding stock and similar declines in their agricultural statistics. In 1920, Alabama reported 782 asses and burros, while South Carolina had 247 asses and burros on its farms.[61]

To improve the overall quality of southern mules, mule producers needed quality stock to serve as the foundation of the program. Locally produced, low-quality animals yielded few profits to breeders. But improving stock was not always simple, and it was never quick. Ideally, breeders needed to follow a two-step process in order to improve their mule stock. Draft stallions mated with mares on hand would produce larger offspring. The mares from these unions and high-quality jacks could then be bred to produce improved mules. The target-size, "market-quality mule" stood at least 15½ hands high and weighed at least twelve hundred pounds. Other observers agreed in principle with improving draft stock but differed on the desired size of the animals produced, with advocates supporting weights from one thousand to fourteen hundred pounds. While debate persisted over the ideal weight for southern mules, observers agreed that the most common mules, which weighed between eight hundred and one thousand pounds, could not properly carry out the tasks farmers demanded of them in order to improve cropping practices, especially in terms of deep plowing. Poor-quality draft stock contributed to low-quality crops and reduced yields per acre.[62] In turn, lightweight, low-quality mules contributed to the region's poverty.

As southern agriculture entered a protracted depression in the 1920s, low work-stock prices created a situation in which many farmers stopped breeding draft animals. The *Progressive Farmer* warned as early as 1920 that the number of horses in mule-producing states was declining, and suggested that southern farmers should, whenever possible, breed their mares to produce a mule or two each year.[63] A few years later, the same journal struck the same theme yet again when it noted that "[s]o far as we can recall there is no other large farming section, equal to the South in area and production, which does not produce more of the horses and mules which it uses. The lack of horse and mule production in the Cotton Belt is peculiar, to say the least." While the author pointed out that the South depended heavily on mules for farming, and that the majority of southerners farmed, he wondered why southern farmers could not raise and feed native mules. Despite the obvious advantages inherent in raising its own draft animals, "the Cotton Belt has failed almost completely in the production of even the horses and mules upon which we are so dependent for the power to do our farm work."[64] G. E. Whitworth, an official with Chicago's Union Stock Yards, warned that "there is a tendency wherever too much machinery or too many barren animals like mules are used to neglect breeding. *The farmers then become like the farmers of the South, raisers of a money crop only.* Farmers of that class always have sharp ups and downs. It is only the diversified farmer that is thoroughly independent."[65] Whitworth was a "horse man," but his comment about the South was pointed—the southern agricultural system rested upon weak foundations. In his view, sharecropping, cotton, and mules combined to form at best a weak region, at worst an area sure to be especially plagued during any economic downturn. A farmer who could not or did not produce his own work animals deserved hard times, and an entire region that did not have the sense to feed itself and produce its own draft animals could expect to suffer hard times.

The downward trend in draft-animal production also worried observers within the United States Department of Agriculture. Among those most cognizant of the draft-stock population, the USDA warned its own employees that draft-horse and draft-mule numbers needed to be replenished through a vigorous breeding program. In mid-1925, the department warned of a one-third reduction in draft horses and a one-fourth reduction in mules by 1929.[66] John O. Williams of the USDA's Bureau of Animal Industry warned that the southeastern states would be particularly hard hit since they had the largest percentage of mules over ten years of age of any region. Williams predicted a constantly dropping number of mules into the 1930s. He pointed out a seventeen percent decline nationally in work animals between 1920 and 1927, and estimated as much as a thirty to forty percent drop by 1932. Not surprisingly, Williams stressed the need for farmers to plan ahead for this "inevitable" work-stock shortage by raising animals of their own.[67] The low supply of draft animals began to affect market prices during the late 1920s. In 1926, horse and mule prices were at a forty-year relative low. By 1928, however, mule prices started to creep upward as the work-stock shortage became more acute.[68]

Rising mule prices encouraged farmers in areas that had traditionally supplied animals to increase work-stock production, but producers found stallions and jacks to be scarce as well as of a high average age. By the eve of the 1930s, horse and mule colt production remained far short of that needed to maintain the population.[69] The draft-animal situation changed little as the nation sank deeper into depression, except that breeding animals continued to age. This further weakened the nation's ability to replenish its work stock. Demand for draft animals kept prices high when other livestock declined in value during the 1930s, but more mares passed the age at which they could produce and care for young.[70]

On the state and local levels, these national trends created a shortage of work stock for farmers. Yet farmers in severe economic straits did not wish to purchase expensive mules to pull farm implements. A large South Carolina mule dealer, commenting in 1930 on the upcoming year, believed his state "weaker in mule power than at any time since the Civil War." For the previous decade, farmers purchased mainly "cheap class mules" and would desire cheap, young mules in the near future, as far as the dealer could tell, although he predicted an increase in demand for older animals as the crop season progressed. Overall, he expected a "slow and draggy" southern mule trade in 1931. Farmers wanted cheap mules during hard times, but once agriculture regained sound footing, experts believed farmers would create a higher demand for quality mules.[71]

The decline of agriculture's fortunes in the face of nationwide depression prompted a renewed interest in local mule-breeding programs, especially since high-quality animals cost more than most southern farmers could afford. In southwestern Mississippi in the mid-1930s, estimates placed the amount of cash flowing out of the region to purchase mules at hundreds of thousands of dollars.[72] The idea behind the local mule-production program also meshed well with what had long been an important focus for extension agents, the "live-at-home" program, which took on new meaning during the 1930s when thousands of southerners relied on their own food supplies and thousands more migrated back to the farms from cities and towns. The *Southern Agriculturist* counseled each farm family in 1935 to "[m]ake yourself as independent as possible of what happens in Washington, Wall Street, or anywhere else in the outside world."[73] An important element of any such program would be the ability to produce draft animals.

To a large extent, midwestern mechanization, by cutting the breeding stock supply, and the Depression, by reducing farm incomes, did what extension agents and agricultural editors could not do. These events created a situation in which farmers could see more sense in producing their own work animals than in purchasing them. New Deal crop-acreage reduction and soil-conservation programs spurred the movement as well, by opening up acreages which farmers could plant with pasture grasses. Farmers and agents joined forces to improve new and existing pastures by draining, liming, fertilizing, and terracing. Extension agents also had for years advocated increased grain production on southern farms.

During the 1930s, farmers possessed less cash than ever with which to purchase mules, but the Farm Security Administration (FSA) made funds available for purchasing mules and jacks.[74] In Arkansas, the FSA and the University of Arkansas cooperated to initiate a mule-breeding program in northwest Arkansas. The cooperative and community-service division of the FSA helped groups of farm families purchase jacks and mares as suitable foundation stock. In some cases, individual farmers purchased jacks after other farmers promised to use the animals' services. In September 1939, the program resulted in the sale of around three hundred colts.[75] Other federal programs helped farmers purchase their own mules. In one study, a picture of a tenant and mule has the caption "Tenant farmer with the mule supplied by the Resettlement Administration, Tupelo Homesteads, Mississippi."[76] Agricultural journals reported similar successes in other states such as Georgia.[77]

Some agents carried out existing work-stock programs with renewed vigor in the midst of the Depression, while others set up new mule-production programs in the mid-1930s. In Covington County, Mississippi, the 1935 extension program included a mule-raising element because of the "scarcity of work stock in the county and the high price of work animals shipped in from the West." The program rested upon two jacks and forty brood mares brought into the county under the auspices of the agent who also gave "special attention" to brood mares at the county fair.[78] Agents often used county fairs and livestock shows as a means of encouraging interest in improved stock of all types. Milton P. Jarnagin of the University of Georgia believed that county and district work-stock shows were invaluable in encouraging local production of draft animals since they displayed actual animals produced on Georgia farms.[79] Farmers inspecting firsthand locally produced animals might be encouraged to undertake such programs themselves.

Mississippi mule production increased gradually from 1935 on and gained such solid footing that one observer expected that over the years an "increasing part of the state's 180,000 mules will be home raised."[80] In southwest Mississippi, an acute shortage of work stock helped agents promote local breeding programs. In Franklin County farmers purchased 180 brood mares, 7 jacks, and 4 stallions in 1935.[81] Itawamba County's agent reported more interest for raising work stock among farmers during 1935 than in the previous several years, while in Smith County, livestock improvement ranked as the "main theme for the year 1936."[82]

In Abbeville, South Carolina, farmers exhibited a "growing sentiment" to obtain mares suitable for carrying mules. During 1935, farmers purchased fifty "good mares." One farmer bought a jack which during the year covered more than fifty mares. In a switch from what many agents had said for years, the agent cautioned that he did not wish the mule-raising program to expand uncontrollably, but hoped many farmers with sufficient pasture would use a mare for field work, breed her, and raise a mule colt. He expected his county, because of its "splendid pastures," to "grow out as good mules as any section."[83]

All local mule-breeding programs did not see marked success. Many agents did not mention mule programs, while others reported little success. In general, however, local work-stock production in the South seems to have increased significantly from around 1935. If a "golden age" of local mule production ever flowered in the South, it was during the late 1930s and early 1940s. Nor were such programs seen simply as temporary expedients until the tractor could supplant draft animals. The *Southern Agriculturist,* hardly the voice of agrarian reactionism, reported a growing interest in colt-raising projects among 4-H club members. Indeed, during the mid-1930s, the journal noted "more interest in breeding livestock and rearing farm work stock than has been for many years." Such programs held particular interest with the older club members, "especially those boys who expect to stay on the farm."[84]

Fig. 5. During the 1930s and 1940s, local production of high-quality mules increased greatly in many parts of the South but was rather short-lived. (South Carolina Cooperative Extension Photographs, Special Collections, Clemson University Libraries)

The Horse Association of America also added its voice to the call for local mule raising in the South during the 1930s. The organization purchased advertising space in periodicals, such as the *Southern Agriculturist,* and pointed out numerous advantages of using home-grown draft animals. The benefits cited included the ability to grow feed for the animals on the farm, the dependability of draft animals, and the small cash outlay needed to use draft animals. The association also argued that farmers who used draft animals instead of tractors supported horse-related industries such as those that produced harnesses, hames, collars, and feed. One advertisement encouraged farmers to "Raise Your Own Farm Power."[85]

While the bulk of extension energy flowed toward white farmers, some black farmers raised their own mules with the aid of black county agents, although the identification of mules with blacks often influenced policies toward blacks. As in many other areas, black farmers received limited resources with which to improve their economic standing. One of the most obvious limits on black farmer participation was the yoke of tenancy. Share tenants retained little control over what they grew, and landowners seldom placed self-sufficiency high on their plans for tenants. Statistically, very few tenants owned jacks and black tenants owned fewer still. According to the 1900 census, "While colored farmers made greater relative use of mules than did white farmers . . . the asses for breeding purposes were mostly in possession of white farmers."[86]

In some cases, prejudice limited the role of black farmers in mule-production programs. Rehabilitation loans or grants directed at whites sometimes included money earmarked for mares, but black farmers usually received mules from government programs. One county agent attributed part of the reason for his area's success with mule raising to the rural rehabilitation corporation which "helped in buying mares instead of mules for white clients." While some government funds flowed to black farmers for the purchase of draft animals, the agent added that those in charge of the funds "felt that most colored clients would ride horses so much that it was best to buy mules for them."[87]

Black county agents may not have held such views of their clients, but more often than not they failed to promote mule raising because they focused on other, more pressing issues. The case of Brown Larson of Sumter County, South Carolina, illustrates to a degree the direction in which black extension agents pointed their clients. Larson, a black farmer, rented a farm for nearly $500 per year. With the agent's guidance and encouragement, the farmer gave up the rental property and purchased a forty-seven-acre farm for $5,500 in 1921 or 1922. Although the buildings needed repair, Larson made do with them and focused instead on improving his crop yields and becoming more self-sufficient. He began keeping records, planted an orchard, and kept purebred hogs, cows, and chickens. He paid off his farm in 1927, and his 1928 crop was free of debt. He then built a new house and barn, and bought a pair of young mules and a horse. Only after Larson achieved a modicum of success, then, did he take the opportunity to purchase stock suitable for breeding to a local jack.[88]

Brown Larson excelled where many farmers, black and white, failed. Although diversification among black farmers has been characterized as "trivial," some black farmers did raise mules and improve their farms in many ways.[89] In

1935 black agents in Mississippi changed little from previous programs except for adding more emphasis to local mule-breeding programs. Black farmers in Rankin County, Mississippi, boasted eleven homegrown mule colts in 1935, and the black agent in Holmes County expected a fifty percent increase in locally produced colts over that of the previous year. Agents noted some successes, but these proved to be exceptions rather than the norm.[90]

As the Depression lingered, southern mule-raising programs became a regular item in the extension repertoire. Extension agents in different areas of the Deep South expressed optimism at the progress of local work-stock programs. Noxubee County, Mississippi, boasted more than 2,000 mares in 1935, of which 563 had been purchased with the guidance of the county agent specifically for the area's work stock–production program. The agent found the "awakening to mule production" very encouraging and expected that within a few years, Noxubee not only would supply its own needs but would export mules to other counties in the state.[91] Mule raising, of course, existed within a larger program of livestock raising promoted by the Department of Agriculture which revolutionized much of the southern landscape after World War II. Improved pastures provided new vigor to cattle raising in particular, but the quality of locally grown mules also improved. J. C. Holton, Mississippi's Commissioner of Agriculture, believed that in 1937 his state produced "more mules and better mules" than at any time in its history.[92]

To some extent, the success or failure of local work-stock programs hinged upon the quality of mules produced. Extension agents stressed the primacy of quality brood-mare stock, as well as blooded jacks, in any draft-animal program. By the late 1930s, agents had convinced a growing number of southern farmers that only by purchasing and using "first-class" breeding stock could they produce quality animals. With quality foundation stock, farmers discovered that they could produce mules superior to those imported from other states. High-quality stock, of course, cost more than poor-quality stock, but once farmers realized the long-term benefits of such investments, they accepted the costs more readily. Thus, cost accounting principles and farm record-keeping programs advocated by county agents helped farmers see the benefits of producing their own high-quality stock. Mule raising, then, fit well with the larger goals of southern extension programs.

In spite of the encouragement from county agents to buy mares and raise mules, the typical southern farmer could not or did not own a jack. Fortunately, counties needed no more than a handful of the animals to breed available mares. Few farmers had the desire, the land, or the capital to purchase and maintain a high-quality jack. B. A. Dickerson of Oconee County, South

Carolina, for example, paid approximately $900 for a jack in 1940. Significantly, Oconee's extension agent pointed out that the county's "leading farmers" acted first to replace old mules with brood mares so that they could raise their own work stock by breeding their mares to jacks such as the one owned by Dickerson. Several county farmers who produced a surplus of mule colts sold them for additional income. The agent concluded that "with the interest in mule colt production that now prevails and inasmuch as the interest is increasing yearly, we will in a few years have a great many home-raised colts" in the county.[93] The services of jacks were also expanded through artificial insemination. As of 1939, for example, 7,100 such community programs had been established in the South. Purebred sires made up more than one-third of the projects. Significantly, the number of high-quality jacks involved ranked second behind bulls. Perhaps even more telling, 589 jacks took part in the projects, compared to 419 stallions.[94]

Unlike the striking geographical variations in the antebellum mule population, a century later both upcountry and low-country South Carolina farmers exhibited increasing interest in home mule production. In low-country Orangeburg County, success bred success. The county agent observed that since several farmers had successfully raised their own mule colts during the previous two years, he expected "a substantial increase in the number of home raised mules" in the near future.[95] In upstate Pickens County, farmers appeared "extremely interested in [the] production of mule and horse colts" and had produced sixteen horse colts and twenty-one mule colts in 1940.[96] Other states took similar strides. In Mississippi, state director of extension E. H. White reported that work-stock production "by farmers has progressed by leaps and bounds during recent years." White explained that community efforts among county agents, bankers, and local leaders had produced these results.[97]

On the county level, breeding-program results could be impressive by the end of the Depression. The Walthall County, Mississippi, work-stock program began in 1935 with the purchase of four jacks and six stallions, which brought the county's total number of sires to six jacks and eight stallions. From 1937 to 1939, farmers purchased 1,050 mares, most of them shipped in by the railroad carload. From this foundation stock, farmers had raised nearly eight hundred head of mule and horse colts by the spring of 1939. Because additional work stock would be needed in the future, and because of the marked success of the work-stock program, the county committee overseeing extension work decided to place "much stress" on work-stock production. Part of the success of the program sprang from simple experience. By breeding draft mares to sires at the proper time, foals would be dropped or born before the spring plowing season demanded hard work from the mares. In this way, brood mares could be utilized almost as efficiently as mules.[98] Approximately one thousand colts under the age of one year lived in Holmes County, Mississippi, in 1940. Unlike most counties, the work-stock program in Holmes had progressed to the point that homegrown draft animals "just about take care of the needs required in making replacements for worn out animals."[99] While not typical, Holmes County's experience in draft-animal raising demonstrated that southern farmers could provide themselves with locally produced work stock.

At the same time, few southern counties could claim independence from outside mule suppliers in spite of the significant trend toward home production of work stock which had developed by the end of the 1930s. In Mississippi counties such as Jefferson, Lee, and Newton, farmers produced hundreds of colts each year during the late 1930s and early 1940s, thus reducing the drain of local capital to mule-producing states. By the latter date, most of the four hundred mares in Jefferson County, Mississippi, were being bred to jacks, prompting the agent to forecast "that soon the annual colt crop will be sufficient to take care of the workstock replacements."[100] Another agent estimated that his county's draft-animal program had reduced mule shipments into the county by approximately sixty percent. Fewer outside animals also reduced the chance of epidemic diseases such as "shipping fever" spreading through an area, an important benefit to all farmers in a veterinary-scarce region.[101] On the state level, one agent estimated that Mississippi could supply itself with around one-half of the twenty-five thousand mules needed each year in the state. This considerable number resulted in substantial savings for farmers since the market price for mules hovered around $120 per head at the time.[102]

Not all Mississippi cotton operations accepted local mule raising as the way of the future. The huge Delta and Pine Land Company never attempted to breed its own mules, despite the fact that the plantation's management had the capital, the room, and the ability. Always keenly aware of the balance sheet, manager Oscar Johnston did not view mule breeding as a profitable enterprise. While Delta and Pine Land Company enjoyed the benefits of scale in many areas, in the case of raising mules for home use the small farmer may have been at a logistical advantage. One or two mares could easily be bred to a local jack. With 1,000 mules on the plantation and 150 new mules needed each year, the difficulties grew substantially. Johnston believed Delta and Pine Land Company's purpose was to produce cotton, not to experiment with mule-breeding programs.[103]

While it is not possible to know whether southern demand for mules could have sustained mule breeding outside the region, it seems improbable. As draft mares disappeared from farms outside the South, millions of farmers who had supplied one or two mules to the South no longer participated in the mule-breeding nexus because their annual crop of a mule or two had been a sideline enterprise. Consequently, little incentive existed, once they had shifted to tractors, to retain mares for the sole purpose of breeding on such a small scale.

Conditions during the war years worked against mule breeding, too, despite a slight but renewed focus on the dwindling supply of mules for the South. The labor shortage hit horse and mule breeding particularly hard. Thousands of farmers who ordinarily took their mares to a local jack to be bred could not spare their own time or that of their hired hands to do so. Because most jacks did not travel well, owners did not as a rule take them around to farms to service mares. They, as well as stallion owners, were also busy answering the call to maximize food and fiber production for the war effort. Consequently, breeding took a backseat during the war.[104] It is also doubtful that large-scale mule-breeding operations could have supplied the South with mules. The *Breeder's Gazette* explored the question and concluded that the "test of time ... indicates that the large scale mule-farm—for one reason or another—is impracticable."[105]

Although local mule-breeding programs were not universal, by the 1940s many county agents had successfully implemented work stock–production programs and could report increased numbers of locally raised mules and a general rise in the quality of draft animals raised. Also by the turn of the decade, farmers who had participated in draft animal–production programs for several years were employing their animals in the field and perhaps selling a few surplus mules, thus encouraging others to take part in the programs. The benefits of local mule production became increasingly clear to participants and observers alike. More counties could boast of tangible accomplishments in each of the three areas necessary in a successful stock-raising program—raising quality work stock for farm use, marketing surplus animals for additional income, and producing an ample supply of homegrown feed supplemented by suitable pasturage.[106] Programs aimed at informing farmers about basic animal care, such as bot control, rounded out the extension agenda.

Mule-production programs made up a segment of a much broader reform movement with roots in the Progressive Era, and county agents carried it throughout the South. While many of the United States Department of Agriculture's programs worked at cross-purposes, in the case of draft-animal raising, many of the broader reform goals such as replacing some cropland with improved pastures, increasing the quality of foundation stock, and employing basic cost accounting and management principles to farming converged to advance local mule production in the Deep South. Mule-raising programs also encouraged farmers to interact with one another even in an era of increasing specialization and isolation.

Indeed, as the traditional rural culture of the South crumbled, in part because of the USDA, the department also advocated programs that encouraged aspects of the old self-sufficient ideal and a community spirit of cooperation. At the same time, tractor numbers increased steadily in the South throughout the 1930s as farmers received payments from the Department of Agriculture's crop-reduction and soil-conservation programs. The same programs that provided capital for tractorization, however, also encouraged stock raising, especially cattle and draft-animal production. Livestock work in the South usually focused upon several types of animals. While beef and dairy cattle, sheep, swine, and work stock all received considerable attention from county agents, beef cattle and work stock remained the primary focus of extension efforts in husbandry.[107]

For decades a variety of southern voices, ranging from agricultural editors to railroad presidents to county agents, sounded the clarion for agricultural diversification and greater self-sufficiency for the region's farms. Nearly a century after the call went out to southerners to produce their own mules, large parts of the South seemed to be developing the ability to produce a substantial portion of their own work animals. At that moment, however, the shift to tractors heralded a revolution in southern agriculture, only to be magnified in the postwar years by the advent of a commercially viable mechanical cotton picker. As a result, the need for draft mules declined precipitously and local mule-breeding programs quickly faded from the extension agenda.

4

Debating Farm Power

The USDA, the Midwest, Mules, and Tractors

Despite the commonly expressed belief that mules would long remain an integral component of the southern agricultural economy, and despite encouraging signs that the animals would be produced increasingly in the Deep South, the mule population in the South plummeted after the Second World War and continued to dwindle during the 1950s. During the early part of the twentieth century, a clash broke out between tractor and draft-animal advocates, each side with its own agenda and each looking out for its own interests, but both driven by the belief that they had the best interests of the American farmers at heart.

No prophet foresaw the rapidity or the extent of change that would occur in the Deep South in the 1940s and 1950s. There had been rumblings of the coming changes in the Midwest, however, where mechanization began earlier than in the cotton South. Yet perhaps because the two regions were starkly different, few of the lessons learned in the Midwest were applied to the South. In addition, as midwestern farmers shifted to tractors, thus reducing their dependence on draft horses, an important source of mules waned. The South's response to mechanization, then, was affected by events in other regions of the nation. Still, as the Corn Belt shifted toward tractors, there was no clear indication that southern draft animals would also disappear within a generation. The leading agricultural journal of the South noted that draft animals "had not been man's useful servant for these thousands of years to get off the face of the earth at the order of any mere inventor or manufacturer of lifeless machines."[1] Similar sentiments were expressed time and again until mules actually disappeared from southern fields.

Agriculture officials confronted the difficult issue of draft animals versus tractors first in the Corn Belt, and in that setting developed attitudes and policies related to tractorization which would later have a profound impact on southern agriculture. Thus, while geographically removed from the South, the shift to tractors in the midwestern states set the stage for how the government viewed southern mechanization, and is crucial to understanding the relatively disinterested attitude the USDA held regarding many of the profound changes southern society underwent during the 1930s, 1940s, and 1950s. This is not to say the USDA was not deeply involved in the South. The federal government poured money and manpower into the South, and such federal largesse usually benefited eager, large landowners and neglected small farmers, tenants, and sharecroppers. It is also important to note that there were differences in how the USDA and other agencies, such as the Bureau of Agricultural Economics, viewed farms and farmers. The point remains, however, that the federal government often expended much of its energy studying *what* was happening, more than trying to *direct* the changes. This may have stemmed from conflicting views within the agencies, as well as the fact that much of the leadership had grown up on farms and had experience with draft animals.[2] Having said that, the role of the USDA in the process of tractorization has not been explored in detail, and historians have yet to explore fully the various factors related to tractorization.

By the time the draft animal–tractor debate engulfed the USDA, the automobile and, to a lesser extent, the truck already were fixtures on many American farms. Rural people across the nation had embraced automobiles by the 1920s.[3] In every state the percentage of farms reporting automobiles was substantially higher that that of farms reporting tractors. On the national level, only 3.6 percent of farms had tractors. The ratio of tractors to farms varied from region to region and crop to crop, but in no region as defined by the census did more than nine percent of farms report tractors. A few states such as North and South Dakota boasted more than fifteen percent of farms with tractors, but this phenomenon was unusual. During the years following World War I, draft animals still served as the main source of plowing and cultivating power. Thus, it was hardly a given that tractors would supplant them in only a few decades, and for that reason draft-animal advocates believed they had a reasonable chance to convince farmers to continue to hitch their plows to animals and not to machines. Most people simply could not conceive that working horses would not have an important place in the modern industrial and agricultural worlds.[4]

While the USDA unflinchingly advocated increased efficiency and greater productivity in agriculture, it also expounded the virtues of self-sufficiency and of the family farm. Over time, however, efficiency and productivity won, and in the end, the "USDA did not simply propagate improved methods—it became the Church of Information and Technology (with its own missionaries) for millions of modernizing farmers. Its experts eventually embraced any machine or chemical that promised increased production regardless of how technological change would affect farm families or the environment."[5] The USDA did not immediately embrace technology without question, nor did department officials always see tensions between the efficiency and self-sufficiency. Officials constantly stated and apparently believed that American farmers could have the best of both worlds. Despite the fact that a conflict had long existed between the family-farm principle and the political exigencies of modern statehood, the USDA insisted that the Jeffersonian and Madisonian ideals could be harmonized on the American farm. It was perhaps a naive view, especially in the southern context, but one genuinely held by agricultural officials who, according to one scholar, had not pushed mechanization hard enough.[6]

More often than not, policies that affected not only southern agriculture but southern rural society developed outside the South. The treatment of tenants under the AAA is perhaps the most obvious example of well-intentioned but ill-fated decisions, yet it is hardly the only one.[7] To be fair, the USDA faced enormous difficulties, not the least of which was the logistical difficulty of implementing policies in a nation as large and agriculturally diverse as the United States. Yet, on apparently clear-cut issues such as the tractorization debate, the department often found it difficult to take and hold a position.

On a broad scale, the Department of Agriculture stressed scientific farming techniques and the use of labor-saving machinery in order to increase efficiency while reducing the man-hours required for specific operations such as harvesting grain or planting a field. The land-grant college system played a key role in implementing the USDA's vision. Driven by "a progressivist fervor that equated progress with science, mechanization, and capital-intensive methods on the farm, in the farm home, and in the factories," the land-grant college and its progeny acted as a prime mover in the technological revolution in agriculture, although the reasons for this have not been fully explored.[8] According to the USDA, greater productivity ensured that farmers would retain an edge in a highly competitive environment. For all of its trumpeting of the value of scientific farming and labor-saving machinery, however, the department had no guidelines for dealing with the rise of the tractor during and immediately following the World War I era. Consequently, the USDA seemed unsure of its ultimate goals as it groped toward firm policies in this area. The stakes being high, a storm of debate broke upon the Department of Agriculture as tractor numbers increased rapidly in the Midwest.

Much of the inability of the USDA to make generalizations beyond seeing a future with both types of power, of course, rested on the local or regional nature of agricultural choices. So, to some degree, the national scope of the

Department of Agriculture itself limited its ability to address the question of horses versus tractors in any but the most general way. But USDA officials did not seem to employ this argument. The state land-grant university, experiment stations, and extension systems acted in a more local context, but the enemies in this battle focused on Washington administrators.

The USDA refused to take a stand during the early stages of the debate between tractor and draft-animal proponents and therefore became a lightning rod for criticism from both sides. While a multitude of interrelated issues swirled within the controversy, the core question was simple—would draft animals or tractors better serve the nation's farmers and future? Certainly there existed powerful currents within the agricultural sector that pushed farmers to mechanize. Agricultural engineers, for example, believed that "if a machine could perform a task, it should." In addition, agricultural engineers "worked hard to create a context in which 'power farming' would replace traditional farming." As far as the agricultural engineering community viewed it, then, machinery was at center stage in America's agricultural future.[9] From today's perspective, the outcome may seem inevitable, and because of that one may assume that the shift to tractors would have appeared as such to contemporaries. That, however, is hardly the case since "the mechanical solution was or is not inevitable at all relative prices."[10]

The USDA sought to understand farm management whether the farm used mules or tractors. In the early-twentieth-century South, however, the efficiency equation did not include tractors. A 1918 United States Department of Agriculture study of an upstate South Carolina county where most farmers grew cotton and plowed with mules found that the mule was "the pivot" around which each farm revolved.[11] The same could be said of practically any cotton farm in the South. Next to man labor, mule labor constituted the largest cost on the farm, larger than the outlay for rent or fertilizer. While man labor cost Belton, South Carolina, farmers nearly 38 cents out of every dollar that went toward producing crops, mule labor cost him 21 cents per dollar. Rent and fertilizer averaged around 16 cents out of each dollar spent on producing a crop.[12] Farm equipment manufacturers estimated that eighty percent of a farm's operating expenses went for man and horse labor.[13] Added together, man and mule labor, fertilizer, and rent totaled more than ninety percent of production costs on Belton area farms in 1918.[14]

A striking aspect of the Belton study is the treatment of the issue of farm size. The author, A. G. Smith, was far more concerned with good management techniques than he was with size per se. His study examined and developed principles with which he believed "anyone operating a farm under the condi-

Fig. 6. The stereotypical southern farmer used mules to plow eroded fields. As this photograph shows, however, mules and good farming were not mutually exclusive. Here mules are being used to terrace a field to prevent erosion. (South Carolina Cooperative Extension Photographs, Special Collections, Clemson University Libraries)

tions that prevail on the farms surveyed may . . . organize and operate his farm with a high probability of success." Conversely, individual farmers and farms might dictate slight changes in the general plan Smith laid out, "but, with rare exceptions, wide deviations from it probably will be followed by reduced returns."[15] Small farms were found to be less profitable than larger operations, but overall the scale of farming remained small. The key organizing principle was how many acres each mule worked on a farm. Beginning at eleven acres per animal, Smith calculated that income per mule rose as acreage per mule rose. While the income per mule was $238 at eleven acres, it reached a peak of $422 per mule at the twenty- to twenty-three-acre mark. As acreage worked per mule continued to climb, the profit per mule dropped to $386 at twenty-eight or more acres.[16]

According to the report, a farm's particular size was not as important as its ability to efficiently utilize its mule labor, since the number of acres per work animal was "an indication of the utilization of the labor and equipment and of the efficiency of the farm organization."[17] The highest return per work animal fell within specific acreage limits, in this case between twenty and twenty-three acres. The most profitable farms could assign acreages within those boundaries to each mule. Clearly, large farms could be more flexible in this than small farms. For example, using the twenty- to twenty-three-acre rule, a Belton farmer with two hundred acres of cropland could employ nine or ten mules. A farmer with only thirty acres, however, found himself at a disadvantage because thirty acres was too much land for one mule, but not enough to justify two mules. In short, the Belton study found that there existed specific sizes of farms that best suited specific numbers of mules.

The thrust of this analysis, known as the law of recurring efficiency, found that size was important, but that it was more important for a farm to be properly sized so as to use its draft power properly. "Handling the farm," the report explained, "in such a way that the unit of organization can be used at its optimum capacity is therefore one of the important factors in determining the success of a farm." The favorably sized farms edged out the unfavorably sized farms in other areas, too. They produced crops at the lowest cost, their dwellings had a higher value, they had a lower cost per man and mule day of work, and their average yields of crops were higher. Good farmers organized and operated their farms more efficiently than poor farmers, even before the USDA stepped in to analyze them and make recommendations.

The report concluded, "When a farmer has the foresight and ability to adjust the size of his farm so that the labor and equipment have a high degree of efficiency, he will also usually have the ability to secure yields that are above the average."[18] The findings of this study, then, would have benefited the less profitable farmers of the Belton area, if they chose to apply the suggestions to their operations. Significantly, the study demonstrated that good farmers organized their operations efficiently and poor farmers did not. In short, some farmers were better at realizing a living from the soil than others, and studies such as this one sought to find what made the good farmers good so that less successful farmers could learn from them.

Size, then, was not the focus of the USDA's study in Belton. Indeed, the author found that two-mule farms, those having forty-one to forty-five acres, were for their size more profitable than one- or three-mule farms. As the author concluded, "Evidently a two-mule farm with the proper acreage per work animal is the best size of farm for this community."[19] The USDA did not focus solely upon large farms as better farms, at least in the South in 1918. As time passed, and certainly as technological advance perfected the tractor, size became more important to farmers and to agriculture officials, but as long as the mule remained the "pivot" of farm organization, small farmers could compete with larger operators. Once the lessons of the Midwest became manifest and were applied to the South, however, size and efficiency did become more synonymous in the USDA lexicon. Tractors unbalanced the efficiency equation, and as an unknown variable, questions quickly arose concerning how they would affect farmers' decisions.

Both tractor manufacturers and the draft-animal industry vied for the attention and blessing of the Department of Agriculture during the early twentieth century; but perhaps because they had the most to lose, horse and mule adherents clamored more loudly than did the tractor industry. No one group or individual raised a more pointed and consistent voice for the continued use of draft animals on farms than Wayne Dinsmore, first as secretary of the American Percheron Society and later as secretary of the Horse Association of America (HAA). The HAA was formed in late 1919 and incorporated on January 15, 1920. In its first eight months, it collected more than $70,000 mainly due to the efforts of its directors.[20] During the 1930s, the association changed its name to the Horse and Mule Association of America, but changed back to its original name in 1948 when the focus shifted to pleasure horses rather than work animals.

Dinsmore waged a relentless battle against anyone who advocated replacing the nation's draft animals with tractors. Never content to conduct a defensive battle, Dinsmore, under the HAA's standard, not only attacked his opponents but also aggressively established his own program designed to convince farmers, first in the Midwest and later in the South, that draft animals made solid economic sense.

Farm power advocates had for some time expressed interest in government-sponsored research of farm power questions. Society president P. S. Rose announced in 1910 that his group was "unanimously" in favor of seeing farm machinery investigations funded through governmental action. Rose felt it "advisable to have a bill introduced in Congress setting aside a special fund for this work." Rose was not averse to his own organization playing a key role in carrying out such work, and while he did not mention his preference, his loyalties were clear enough. He corresponded on the letterhead of *Gas Review: A Magazine for the Gasoline Engine User.*[21]

Despite their deep differences, there was one issue on which the draft and power farming advocates agreed. Both sides encouraged the Department of Agriculture to undertake investigations into the farm power question. The American Society of Agricultural Engineers, for example, early in the twentieth century pushed the department to examine the matter, as well as suggested that it investigate broader issues related to farm machinery. This group submitted to Secretary of Agriculture David Houston a proposal covering the need for and scope of a bureau of agricultural engineering within the USDA.[22]

In the midst of the growing debate between tractor and draft-animal advocates, USDA officials moved to schedule a "Farm Power Conference," which met in Chicago in October 1919. Representatives from the USDA, state colleges, and numerous trade and agricultural associations comprised the more than thirty committee members at the conference.[23] As the conference date approached, a clear rift developed between "tractor men" and "horse men." Feelings ran high enough so that each group met separately the morning of the first day before convening jointly in the afternoon. The committee recommended that seven projects be undertaken. The proposed studies covered both sides of the draft animal–tractor debate, but several of the studies sought to standardize how draft animals and tractors could be compared. The recommended studies included: testing and rating farm tractors, determining the working rate of horses, measuring power requirements of machines and implements, developing practical methods of expanding the power of farm horses, determining the mechanical efficiency of horses as power units, increasing the economic efficiency of horses and tractor power by adjusting the size of the farm and combination of enterprises, and compiling accurate data concerning farm power demands and the relative cost of meeting these demands by the various kinds of power on farms.[24]

Both sides of the debate received consideration, but the meeting may have prompted the draft-animal industry to organize and develop countermeasures. One of the committee members was Wayne Dinsmore, then secretary of the Percheron Society of America. It hardly seems coincidental that the Horse Association of America, with Dinsmore as secretary, formed soon after this conference adjourned. Even so, draft-animal proponents seemed somewhat satisfied with the proceedings, for early in 1920 Dinsmore urged the secretary of agriculture to move ahead to secure funding for the studies proposed at the Chicago conference.[25]

Leaders from the horse world and allied industries created the Horse Association of America as a result of the significant inroads made by tractors onto American farms during the World War I era, but an important catalyst for the organization's zeal was that the Farm Power Conference itself revealed the strength of the tractorization proponents. Supporters included veterinarians, farriers, breed associations, and grain marketers. The economic power and importance of animal-related industries were still significant, although declining in numbers and strength. Tanning, harness and saddle making, butchering, blacksmithing, and other industries tied to livestock and work animals employed one out of every eleven factory workers and tradesmen in 1890.[26]

At its first annual meeting in 1920, HAA secretary Dinsmore observed that horsemen and horse-related industries had "been living in a fool's paradise with respect to the horse displacement which has been occurring" over the past decade. Dinsmore explained that draft-animal adherents blindly believed American farmers would never forsake horses and mules. As a result, they did nothing to counter the information disseminated by power farming forces. "So the manufacturers and dealers," according to Dinsmore, "in other types of motive power have had an absolute clear field, and have been able to make claims that their particular types of motive power were more efficient and much more economical than horses and mules. They have been able to make the claim both in the city and the country, and in the absence of any proof or any information to the contrary, they have been able to get away with it."[27] In his statement, Dinsmore detailed his vision of the new organization's primary purpose—educating and informing farmers about the merits of farming with draft animals.

Dinsmore carried out his task vigorously, if not always diplomatically. To Dinsmore, grim times for the draft-animal industry demanded bold measures to save the horse and mule, the American farmer, and the nation itself from the tractor and, to a lesser extent, the automobile and motor truck. Nor was Dinsmore's a lone voice. The president of the Horse Association of America wrote with decided conviction that "the extravagant use of motor transportation in this country has passed far beyond the point of sanity and unless something is done to curtail this needless waste the ultimate result cannot be other than disastrous for the future welfare of the nation."[28] Hardly self-seeking reactionaries, the HAA officially recognized the worth of the automobile. For example, in 1925 Dinsmore praised the automobile because it had "contributed greatly to equestrian sports, [by] making riding clubs, polo fields and hunt clubs easy of access." In regard to the tractor, however, the HAA's leadership sounded so urgent because they genuinely believed the American farmer was being misled by the tractor industry, which in the 1920s was often the case.[29]

According to HAA president Fred Williams, tractor propagandists had consistently overstated the ability, range, and utility of their products while understating their costs, limitations, and shortcomings. At the same time, they had inflated horse costs, deprecated the horse's utility, and both forecast and urged the horse's "early and complete replacement by mechanical power." Williams admitted the misinformation was sometimes unintentional, but he added that "in the main the propaganda has been deliberate, carefully planned, adequately financed and tremendously aided by the failure of those who believed in the horse to bring forward any facts in their possession or to seek new ones with which to refute the statements of the motor interests." Williams, while harshly critical of the motor interests for their methods, did not take a Luddite tack. "Anyone who isn't blind or a fool," Williams stated, "will acknowledge the value of the motor vehicle and grant it a permanent place *in its proper place.*"[30]

At approximately the same time, secretary of agriculture E. T. Meredith, who served for approximately one year, wrote that he "was very much interested in the utilization of mechanical power and all the other mechanical problems which are related to agriculture [and that] . . . every available worker should have his efficiency developed to the fullest possible extent by the utilization of mechanical equipment wherever it is economically feasible."[31] While the USDA had yet to determine a firm policy in 1920, Meredith's phrase "wherever it is economically feasible" came very close to expressing the policy settled upon by the department. In short, the USDA ultimately took the position that some farmers would benefit from using tractors while other farmers would be best served by draft animals, and this stance continued well into the twentieth century as cotton farmers confronted mechanization. Ambiguity may have been the most pragmatic avenue for an agency charged to serve all the nation's farmers, but it led to many misunderstandings.

While the opponents prepared for further clashes, the USDA seemed unprepared to take sides. The department justified not taking a position by stating that no information existed on which to base a judgment. Secretary of agriculture Henry C. Wallace explained that the "effort of the Department in its research work on the problem of farm power is to gather the facts without regard to whether they are favorable to the horse or to the tractor, and let these facts speak for themselves."[32] Until such scientifically gathered information was available, continued the argument, the department hedged in its responses to the growing number of requests for guidance from farmers after World War I. The USDA preferred to answer inquiries with pamphlets since they were cost effective and usually satisfactory, but in the case of farm power choices it could offer little prepared information to interested individuals or organizations.[33]

H. C. Taylor, chief of the Office of Farm Management, stated in 1920 that while tractor studies were under way, more information needed to be gathered before the department could support tractors. Taylor explained that to endorse "the tractor for all kinds of places might do a very great deal of harm." Over the long term, Taylor feared that a rapid and ill-advised shift to tractors could leave the nation with a shortage of draft animals, a problem that loomed as a real possibility as the 1920s progressed. Taylor also echoed two arguments that horse proponents often employed. First, he noted the signal changes set in motion when a farmer purchased a tractor. Tractor farmers immediately became "dependent upon the city with all its labor problems for his machinery and for fuel and oil consumed in producing the power." Taylor noted that while tractors had their place on some farms, the "typical sales agent is not interested in knowing just where this place is as in selling the tractor."[34]

Many of the confrontations stemmed from the fact that both horse and tractor proponents eyed the department so closely, each jealously accusatory when it suspected the other of misreading or consciously misusing department information to support its own cause. This occurred repeatedly in spite of assurances such as that from Secretary Wallace that the USDA intended only "to gather the facts without regard to whether they are favorable to the horse or to the tractor." Wallace expected the facts to "speak for themselves."[35] Facts could be emphasized, neglected, or twisted, however, and Wayne Dinsmore believed it his prerogative to act as the department's ex officio conscience. For example, he reminded Secretary Wallace before a speech to the Farm Economics Association that the discourse "would not be complete without due heed to the important part which the intelligent use of horses plays in reducing production costs." Dinsmore hoped Wallace would "take time to place some emphasis" on this point.[36] Significantly, both tractor and horse proponents acknowledged the decisive role that the USDA played in the issue, and equally important is the point that tractor interests lobbied as hard as horse proponents to receive the department's blessing.

In the absence of a firm policy line and information, organizations on both sides of the farm power issue mustered their own resources and broadcast their own information. In a typical leaflet, the HAA forcefully set forth the pro-mule position under the rubric "A mule is the only fool proof tractor ever built." Using the Miller Brothers Ranch in Oklahoma, described as the largest diversified farm and ranch in the United States, the pamphlet noted that the ranch had compared tractors to mules through close record keeping and had found that tractors cost three times as much as mules to own and operate under actual farming conditions. A spokesman for the farm left little doubt concerning his position: "We believe all tractors are bad, only some are worse than others. When it comes down to actual facts in dollars and cents, we believe that any farmer who disposes of his horses and intends to do all of his farmwork with tractors, will eventually 'hit the rocks,' and that he is only working for the man who sells tractors, for as soon as he has made enough wheat or other farm products to pay for his tractor, it will be necessary for him to purchase another."[37]

Draft-animal proponents tapped a deep well of distrust that had produced earlier agrarian movements against corporate power. Further, grain and hay crops could be produced on the farm unlike fuel and lubricants for tractors.

Draft animals also consumed large amounts of hay and grain, providing a market for farm products. One estimate stated that draft animals normally consumed sixty-eight percent of the nation's oat crop, forty-five percent of the hay crop, twenty-five percent of its rye crop, and twenty-four percent of its corn crop.[38] Without draft animals in full use, allied industries, including harness makers, blanket manufacturers, feed businesses, veterinarians, horseshoe manufacturers, and farriers, faced hard times as well. The combined arguments were compelling, but only if one shared the same set of assumptions about how American agriculture should operate.

Some observers found the ongoing debate tedious at best. The *Progressive Farmer* seemed fed up with the issue. In a 1920 article critical of both sides in the debate, the magazine pointed out, "If much of the energy consumed in spreading propaganda in support of the horse and the truck and tractor were expended in the development of a better horse, a better truck and a better tractor, the effort would be worth more in results." At this point, the magazine viewed the position of both the truck and tractor on farms as "secure." Even so, it warned that not everyone should purchase them.[39] As time passed, the *Progressive Farmer* became more open in its support of tractors and related farm machinery that could increase a farmer's efficiency. At the same time, the magazine did not dismiss draft animals entirely until mules had all but disappeared.

In 1921 and 1922 the Department of Agriculture found itself in the midst of what one official called "a war to the finish between the power farming people and the horse champions."[40] A pugnacious Wayne Dinsmore, speaking as secretary of the HAA, squared off against an equally combative Finley Mount, chairman of the Tractor and Thresher Department of the National Association of Farm Equipment Manufacturers. H. C. Taylor, the USDA's chief of Bureau of Markets and Crop Estimates, unintentionally sparked the quarrel during a speech he gave before the November 30, 1921, meeting of the HAA. Taylor, speaking in place of Secretary Wallace, referred to but did not directly cite a study of 286 Ohio, Indiana, and Illinois farms. The report concluded that due to high feed costs in 1920, tractors were cheaper to operate than draft animals, while in the following year, horses had been less expensive to operate. Another survey of farms in central Illinois came out in favor of draft animals as well. Among other points, the authors noted that on a farm growing less than 240 acres of crops, tractor use was probably not economically warranted. The study concluded that "little saving in man labor was effected by the use of the tractor."[41]

Dinsmore lost little time in preparing a press bulletin using Taylor's remarks which he released on December 7, 1921. Much to the tractor lobby's chagrin, the bulletin reported that "big national authorities stood up in the Chicago convention of the Horse Association of America last week, and unreservedly endorsed the horse as the logical power for the farmer and the most economical hauling power for city freighting."[42]

Power farming advocates took exception to a USDA spokesman assuming such a position, if he had actually done that. Finley Mount went to the source of the remarks to inquire about the matter. Taylor was not apologetic, but he did explain the context of his words and charged any wrongdoing to Dinsmore's interpretation, not to the speech itself. Taylor's speech focused on the outlook for the next few years for breeders of draft horses, racehorses, jacks, and mules. He gave Mount two reasons why he believed draft animals "should be able to hold an important place on American farms." Taylor noted their adaptability to "every variety" of farmwork. He also believed the argument was strong that draft animals could be produced on the farm and fed with farm products. In closing his reply to Mount, Taylor insisted that he did not say what Dinsmore had reported. "In other words," he concluded that Dinsmore's release "is his own interpretation of the results of the investigations, not my statement."[43]

Mount, dissatisfied with Taylor's reply, appealed to Secretary Wallace that Taylor's letter was "unsatisfactory."[44] The advertising manager for International Harvester supported Mount's offensive. F. W. Heiskel penned a long letter to Mount, which Mount forwarded to Wallace and Taylor, expressing his distress about the USDA's apparent favoritism. Mount testily wrote, "So long as this negative program is carried on entirely by those having a financial interest in continuing the use of horses, the tractor manufacturers can well ignore the movement, but this becomes more difficult when officials of the Federal Department of Agriculture, who have such a wonderful opportunity to be of real service to American farmers and the country as a whole by taking a fair and unbiased position on this subject and treat both sides strictly on their merits, apparently lend their influence to the Horse Association."[45]

Mount followed his strongly worded letter with a more conciliatory message written on Valentine's Day. Most upsetting to the power farming forces was the USDA's unwillingness to condemn Dinsmore's widely circulated press release. Mount observed that Taylor had admitted the bulletin misrepresented his statements, yet Department of Agriculture officials refused to pursue the matter. "What the tractor industry feels it is entitled to ask," wrote Mount, "is this: Will you Mr. Secretary, permit this thing to go unchallenged in this manner?" Mount followed his letter with a personal visit to the secretary's office, but Wallace was not in at the time.[46]

While busy indicting the horse interests for misstating the USDA's position on tractors and draft animals, the tractor industry also took the opportunity to exploit any openings. Tractor proponents seized upon a telegram sent by Secretary Wallace to the St. Paul, Minnesota, *Pioneer Press* regarding the farm power question. Wallace stated in part that the "Department of Agriculture is making careful study of this power question because we appreciate the need of making the largest possible use of mechanical power which reduces production cost."[47] Working from the assumption that mechanical power always represented reduced production costs, which the USDA had not determined, the *Power Farming Press* surely took some pleasure in observing that the "Department must be thoroughly sold on the power-farming idea to express so specific an opinion. This is gratifying to the tractor and power-farming equipment manufacturers, to say the least. Moreover, it is what *Power Farming* has been urging for years."[48]

Wayne Dinsmore expressed concern over the misapplication of Wallace's statement, but assured the secretary that his telegram "was entirely O.K. for it is quite desirable to use mechanical motive power which reduces production costs. Unfortunately, no such mechanical motive power exists and the construction which the *Power Farming Press* put on your telegram is far from the interpretation you yourself would put on it."[49] Wallace replied to Dinsmore in a confidential letter in which he stated, "If I stay here much longer I shall be suspicious of everybody and everything. I am beginning to understand why many people in government service seem afraid to open their mouths." A weary Wallace, at that time attempting to gather a consensus among his staff on how to close the matter, penned what sounded much like a departmental position statement: "Both the horse and mechanical power have places on the farm, and will find their places, whatever may be said or done by the zealous, and perhaps at times over-zealous, parties of either." He closed by expressing doubt that "discussions" of the type in which the department had been involved served any useful purpose.[50]

Throughout February, March, and April, Wallace attempted to hammer out a viable policy statement guided by experts on his staff. In the end, the department fell back upon its original plan to say as little as possible while conducting studies. A typical policy statement contained several major points. First, it noted that "both the horse and the tractor have a place in American agriculture." Perhaps hopefully, perhaps naively, the statement observed that "when each finds its place, there will be no conflict of interest between them." Of the statement, one staffer commented that "it is about what we should say when it is absolutely necessary to say something."[51] The USDA believed that over the long term some farmers would continue to use draft animals, while others would end up as tractor farmers.

The department surmised correctly, as later information supported, that general rules were not easily developed, but at the same time it set out "to ascertain the truth as rapidly as possible."[52] In its treatment of tractorization, the department tended to react to changes by studying what was happening or had happened on farms, rather than developing specific, overt policies to shape the use of draft animals or tractors in a particular direction. At the same time, the department's emphasis on increasing output per man-hour and per acre played a crucial role in how it viewed the horse and tractor confrontation. As one draft-horse advocate and critic of the USDA has commented, "Measurements of efficiency in food production . . . on the basis solely of human labor expended per lb. or acre are invalid and misleading."[53]

At the time the mule and tractor controversy shifted to the southern United States, the USDA was still content to gather information and let the horse and tractor proponents fight their own battles. Neither side left the department alone after 1922, but by the same token, neither side expected the department, as both had hoped in 1922, to openly endorse one position. The conflict shifted more to the pages of trade journals and agricultural periodicals. Secretary Wallace was not unmindful of the potential repercussions of the internal combustion engine upon agriculture, but he desired quantifiable data instead of impressionistic observations. In a philosophical inquiry, Wallace wrote to H. C. Taylor, chief of the newly created Bureau of Agricultural Economics, regarding the changes in agriculture since the turn of the century. "Twenty years ago," he noted, "the power on the farm was furnished mostly by horses and windmills. The horses were fed on stuff grown on the farm and this did not require an outlay of cash. Wind was free. All we had to do was to stick a wheel up in the air and harvest it. Now a large part of the power is furnished through engines. We have to pay cash for gasoline and oil and for repairs. Grain and hay formerly eaten by horses is now sold. Have we any way of measuring the effect of this change?" Wallace also alluded to societal changes that had occurred. "Twenty years ago," he mused, "there was a good deal more swapping of work by the farmers of various communities than there is today. Nowadays not so many care to exchange work. Have we any way of measuring the additional cash outlay because of this change[?]"[54] The department developed methods to measure change. But the USDA did not concern itself at this point with potential repercussions that might be wrought by tractors and other machinery.

Large tractors could clearly pull larger and heavier implements than the small teams of horses or mules most farmers used. Draft-animal forces, led by Wayne Dinsmore, responded to tractor claims of greater efficiency by advocating that farmers hitch larger teams to heavier farm implements. This produced an ideal result in the view of draft-animal proponents—greater efficiency and increased use of draft animals. The HAA took action to educate American farmers on the benefits of the "big hitch." The organization implemented a paid advertising campaign in the late 1920s, as well as undertaking its own study into the costs of producing, raising, maintaining, and employing horses and mules. In 1929, for example, the HAA mailed more than 106,000 pieces of printed matter. It also received approximately nine hundred cards and nearly sixteen thousand letters. Its publicity campaign boasted 13,119 inches of type in magazines and newspapers with a combined circulation of almost thirty-six thousand.[55] The big-hitch campaign information also reached many southern farmers through the pages of the *Progressive Farmer,* which summarized the program and gave the address for securing the HAA's booklet on the topic.[56]

Much of the association's energy and finances went into producing literature advocating larger teams and carrying out demonstrations of large hitches of draft animals. Not surprisingly, early efforts focused on areas where large teams of animals could be most easily adopted. Big-hitch demonstrations numbered 109 in South Dakota, 380 in Iowa, and 142 in Wisconsin during 1927 and 1928. Southern states also witnessed some demonstrations during 1928. South Carolina hosted four, Georgia seven, Alabama two, Tennessee two, Missouri three, and Texas thirty-three big-hitch demonstrations.[57] The HAA carried on its southern work the next year as well.[58]

The demonstration program attempted with some success to utilize agricultural extension agents to aid its design. For example, Dinsmore distributed an HAA leaflet, *Keeping Farm Teams at Low Cost,* to every county agent in the United States. In addition, he sent fifty-five copies to the USDA's Washington office to be distributed among extension staffers there.[59] Dinsmore tested the patience of Department of Agriculture officials at times, however, which resulted in an occasional rebuff. Oklahoma's state extension director, D. P. Trent, refused to order his staff to participate in HAA big-hitch demonstrations, allegedly because extension programs kept them too busy.

Dinsmore penned an ungracious letter to C. W. Warburton, national director of extension work, in Washington. Dinsmore asked that Warburton "put some pressure on Trent" to cooperate with the HAA. "For some unknown reason," he continued, Trent "seems to consider it more important to have men go out and show men how to feed cattle than have them go out and show men how to use their teams in such a fashion as to save from $200 to $300 per year, and I am getting weary of it." Dinsmore then posited that Trent's personal indignation at not having been contacted in person by Dinsmore might explain his reluctance to cooperate. He concluded in an equally presumptuous fashion, "Anything you can do to line him up will be greatly appreciated, but early action is necessary."[60]

Warburton's final reply, although less combative than his original draft, retained a decided edge. He first explained that the federal government would not dictate state-level extension programs, and should it attempt to do so, "that dictation would very promptly and very properly be resented." Warburton then proceeded to defend Trent and to justify his actions. Taking Dinsmore to task in no uncertain terms, Warburton concluded, "if I was in Director Trent's place I doubt very seriously whether your letter of March 21 would influence me to do the thing you desire. I am quite sure, in fact, that this letter . . . would make me feel that you were rather officiously trying to dictate my program, and I think I would resent your attitude."[61] Dinsmore's reply was curt. He said simply, "I have your letter of March 28th relative to the Oklahoma situation. Apparently we will have to proceed through other channels."[62]

In spite of episodes in which Dinsmore did not get his way, the working relationship between the HAA and most extension programs usually proved more congenial and productive. Whereas agricultural colleges and extension departments might have been difficult initially to convince, by 1929 Dinsmore could report a significant improvement. He believed that the "agricultural colleges and their extension departments are working with us. They don't depreciate the tractor, but at the same time they push our work [and] spread our knowledge."[63] Nor were Dinsmore's efforts without fruit. Livestock specialist W. R. Hauser noted that several South Dakota farmers were working eight, nine, ten, and twelve horse teams, and doing more work than two or three men had before. Hauser also called the big-hitch demonstration the "best and most interesting and also the most practical" demonstration he had witnessed in fifteen years of extension work.[64]

Editors of the annual *Yearbook of Agriculture* included a favorable article on big hitches in 1927, further proof that Dinsmore's gospel reached many, both in the fields and in Washington. The article's author mentioned the HAA by name, praised its efforts in cooperation with extension agents, and concluded that big hitches "point a way to a practical way of effectively reducing operating costs, yet retaining on the farm the reliable form of drawbar power which is self-replacing, consumes home-grown fuel, and has the maximum flexibility."[65]

The big-hitch movement, although of restricted usefulness in the South, did echo in the region during the 1930s when a push occurred to convince farmers to use larger mules and one-row equipment.

Despite its successes, the big-hitch program failed to stem the rising tide of tractorization and hardly made any difference in the South, outside isolated areas suited to large teams. While few large teams plowed southern cotton fields, the big-hitch program and the USDA's stress on efficiency did encourage many southern farmers to switch to two-mule equipment. In South Carolina, for example, one extension official noted in 1930 that the state's agriculture was going to center, "for a time at least, around this two mule machinery."[66] Many of the South's farmers relied upon a single mule for field work and could not dream of using large teams in order to improve efficiency even at the end of the mule era. Still, the example of the HAA is important because it represented an alternative to tractorized American farms and it made its voice heard in many areas of agriculture, the farm, the USDA, and even in engineering circles.

Dinsmore himself belonged to the American Society of Agricultural Engineers and on at least one occasion presented a paper at its annual convention. For his perennial optimism about the place of draft animals on farms of the future, Dinsmore must have understood the daunting odds against his view of the future. For example, he chastised the ASAE for focusing on one side of the issue, especially when tax money funded research. Dinsmore reminded state-employed engineers that they owed taxpayers "the direct duty of increasing, in every possible way, the efficiency of motive power used on their farms, and this duty should take precedence over work designed to bring out and popularize" tractors and other types of machines.[67]

Yet, the HAA's big-hitch program contained a fatal flaw from the beginning. Although the differences in employing draft animals and tractors are profound, as most observers noted, the HAA's leadership and that of its opponents were not totally divorced on philosophical grounds. In short, both believed that bigger and faster were better. Dinsmore and the HAA leadership chose to confront tractor forces largely on their own ground by defining efficiency in terms of acres plowed or planted per day or per man-hour. And in this context the big-hitch program was inaugurated. It was a bold stroke and took the battle to the enemy.

In the early years of tractor development, this approach may have made sense. As tractor designers developed smaller, more powerful, and more reliable units, however, draft animals lost the field in terms of pure pulling power. Farmers simply could plow more acres per day with tractors. Tractors could also produce more horsepower per day, particularly after electrical lighting permitted night work. As larger engines were developed, the pure muscle of tractors vastly outclassed that of even large teams of draft animals. Draft animals still held certain advantages, such as the ability to get into wet fields sooner than tractors, but the definition of good farming came more and more to be measured quantitatively—acres covered per day—and with that change the farm horse and mule were doomed. Southern farmers as a whole took longer to convince of this fact, but in time they too accepted the USDA's and the tractor industry's definition of efficiency and good farming, or they left the land. Certainly this definition became one of the most common standards employed in extension studies, although it was not the only one. In a sense, then, the battle waged by Dinsmore and the HAA, especially the big-hitch program, helped prepare the South for tractors.

While powerful assumptions existed about science and engineering within the USDA, and while the USDA as an institution advocated scientific farming and the use of labor-saving machinery whenever possible, the department was also made up of men who held certain misgivings about tractorization. It is clear that there was some room in the USDA for differing opinions on specific issues, at least until mechanization became an accomplished fact, or nearly so. At that point, the old guard had either retired or had learned that advocating its position within the department would do no good, but vestiges remained into the 1950s.[68]

The pro–draft animal feeling took two forms. On one hand, some USDA officials stood unconvinced that tractor farming was universally as profitable or efficient as horse farming. Efficiency rang as a central theme of USDA rhetoric, but during the early twentieth century, and until the post–World War II era in large parts of the cotton South, tractorization and efficiency did not become fully synonymous. In the end, what may be most surprising about tractorization is how little the USDA really concerned itself with overtly advocating a shift to tractors during the early stages of the debate. Historian Anthony Badger has concluded that governmental intervention, especially in the form of the agricultural New Deal, "was marginal to the major developments—the flight from the land, mechanisation, and the consolidation of small farms into larger holdings—that were to revolutionise American agriculture in the next 50 years."[69] In the end, implement manufacturers and dealers pushed tractors much harder than did any government agency.

The Department of Agriculture remained acutely aware of the rising number of tractors, and the consequent decline of draft animals, through agricultural census returns, reminders from horse and mule advocates, and letters from interested or concerned individuals across the nation. Farmers in touch with local conditions perceived more clearly than officials in Washington the implications of tractorization, especially as the Depression tightened its grip on the nation's rural areas. One South Dakota farmer struck a theme later echoed by southern observers of tractorization. "Tractors cause the [*sic*] over production and it encourages corporation farming, which is a great drawback for agriculture. It crowds the small farmer and the beginner out of the game."[70] In addition, to many observers, tractors led to massive displacement of farm laborers and tenants from the land.

But USDA spokesmen insisted that their agency must view agriculture from a national perspective. One agriculture official responded to a complaint regarding the tractor's impact on local conditions by admitting that local and specific examples might point out problems with mechanization, but that "the matter can only be considered on a larger scale or national basis."[71] Many observers linked the rapid rise of tractor use in many parts of the nation during the 1920s with the onset of the Depression. The most vigorous critics of the tractor reflected strong reactionary tendencies. A Texas doctor asserted in a letter to President Franklin Roosevelt that "[a]ll tractors and trucks must be destroyed."[72] A Nebraska farmer's proposal went a step further. He believed that the solution to the Depression was "to do away with tractor manufacturing."[73] Other suggestions sounded less reactionary but aimed at reducing tractor use nonetheless. An Ohio farmer, for example, suggested the government impose a tractor tax punitive enough to curtail tractor purchases.[74]

Not surprisingly, Wayne Dinsmore of the HAA offered his analysis and recommendations to the USDA as tractor numbers increased. He pushed for educational programs designed to give horses and mules equal billing with tractors. Dinsmore acknowledged to Secretary Wallace the USDA's inability to "compel farmers . . . to put horses or mules in place of trucks and tractors as they wear out." He did ask the department, however, to point out the benefits of draft animals, especially in terms of consuming excess grain produced on farms.[75] Voices within the USDA complex also expressed concern over farmers unwisely shifting to tractors because of "hard sell" techniques by tractor dealers such as the purchase of a farmer's draft animals at above-market prices to induce farmers to buy tractors. In reality, this meant that tractor dealers offered a high trade-in value for draft animals. Credit played a role since tractors could be bought on "very liberal credit, sometimes too liberal." Dealers also used a serious outbreak of equine sleeping sickness to undermine farmer confidence in draft animals. As a result, some observers called for a governmental subsidy to stabilize the draft-animal industry.[76] During the 1930s, much of the nation's attention turned to the South, where the peculiar social and agricultural systems made both draft-animal and tenant displacement more visible. Agricultural economist Paul Taylor toured the South in 1937 to gather information on the plight of farm tenants, some of whom had been displaced due to farm mechanization. He reported a significant amount of tractor buying in Texas, where conditions such as large fields suited mechanization.[77]

Over time draft-animal advocates within the department disappeared. Even in the face of tenant displacement or the wholesale replacement of horses and mules on farms, the department refused to intervene in the farmer's decision-making process in the matter of farm power. The USDA did not wish to "do anything which might make it harder for the farmer to cultivate his soil." Crop-control programs clearly intervened in the sense that they limited a farmer's planted acreage, but officials felt it best to let the farmer "be the judge as to the equipment he uses."[78] This attitude remained constant throughout the years after the initiation of crop-reduction programs. In 1941 the secretary of the department echoed this view: "We feel that farmers should be free to choose the type of power that they use, and that programs of the Department should be formulated so as not to jeopardize their freedom, either directly or indirectly."[79]

While the USDA did not feel it should dictate what equipment farmers chose, it was involved in promoting power farming in other ways. As early as the World War I era, for example, Virginia Polytechnic Institute hosted power farming demonstrations in conjunction with several equipment manufacturers. In 1919 seven plowing demonstrations were held with more than thirty thousand farmers in attendance. More than ten thousand farmers were reported at one demonstration alone.[80] In 1922, the school held six power demonstrations at which "tractor and modern power implements were shown in operation doing various kinds of work." Demonstrations of only tractor plowing were discontinued, however, since officials believed that such focused exhibitions had served their purpose in the state. Thousands of farmers attended the more broadly conceived power farming exhibitions.

The decline in plowing demonstrations probably reflected a general trend in research. During the early 1920s, farm machinery projects fell on hard times as agricultural engineering departments found that funding for indefinite or general projects did not materialize as it had in the past. Instead, narrow projects with carefully considered purposes and methods were much more likely to receive funding. In Alabama, for example, the state experiment station narrowed its tractor study to a project concerned with the relationship between tractor wheel slip in the soil and plowing efficiency.[81]

The Department of Agriculture more often than not found itself reacting to changes catalyzed by technological advances. Despite the drawbacks to such a defensive approach, without the ability to dictate fundamental decisions to farmers, the USDA found its policies often one step behind the progress it often encouraged. "Every time a new step in mechanization is taken on a farm," an undersecretary wrote, "a whole train of effects is started off, some good and some bad—and it is our job to help the farmers meet the bad effects."[82] The USDA also worked from the assumption that farmers bought and used tractors because they liked them. Mordecai Ezekiel, an economist with the USDA, summed up several key points when he explained that farmers "like tractors because of the saving time in their use and the reduced nuisance of caring for horses throughout the year, as well as for their greater economy on the job."[83]

Such comments invited criticism, as they sprang from bureaucrats whom Wayne Dinsmore branded in a letter to USDA secretary Arthur M. Hyde as "motor-minded" officials. He complained as early as 1930 that the department was filled with individuals biased toward mechanization in general and tractorization in particular. Dinsmore himself heartily supported more efficient farming operations, but he attacked the prejudice against draft animals he found in the USDA. Referring to the Bureau of Agricultural Economics, Dinsmore noted that the "qualities that lead men into that field of work,—the scholarly tastes, studious mind, inclination to spend long hours in an office studying figures, charts, maps, graphs and accounts—records of things that have passed—make them mechanically minded."[84] USDA scientists and bureaucrats, Dinsmore implied, did not understand how the real world of farming operated.

Dinsmore also called attention to individuals within the USDA who had expressed to him "the view that farmers were going to tractorize, wholly or in part, even though it added to their expenses and reduced prices for farm products—that the attempt to obtain ease, to escape chores, would cause a general shift, no matter how detrimental it might prove to farm interests as a whole."[85] Years later, while speaking at the Horse and Mule Association's annual meeting, Dinsmore admitted that the view of many individuals in the USDA that tractorization was a foregone conclusion was correct. Optimistic as usual, he nevertheless warned that the crucial factor working against the farm draft animal "is the inclination of the human race to be lazy. We are all alike as far as that goes, and a very large proportion of farmers and farm boys are just too lazy to take care of four good horses or mules, or six of them, to get them up and groom them and take care of them. They want the easiest way out. They want to get their work done in less time."[86]

Another strong factor in the tractorization equation, which Dinsmore hit upon in his letter to Hyde, was the idea that tractorization, for all of its revolutionary implications, and perhaps partly because of them, meshed with the nation's technological evolution. In 1920 Secretary Meredith explained that technological innovation in agriculture "may be regarded as an inevitable result of the progress of modern industrialism."[87] As such, the shift from draft animals to tractors had to be faced, not as a boon or threat, but as a fact of modern agricultural practice. Secretary of agriculture Ezra Taft Benson expressed in the 1950s his own "regret" over the decline of draft animals, but like Meredith believed that it had been "inevitable." He further noted that the sense of loss he and his generation felt so keenly "was not so prevalent among the younger generation."[88] Even before USDA policies began to alter the nation's, and especially the South's, agricultural landscape, Secretary Hyde called attention to the fact that the "gas driven vehicle had a vast reaction upon the farm market." Evidently dismissing critics such as the HAA's Dinsmore, he commented that "[n]obody is critical of this for the reason that it is merely one of those modern economic shifts which will always occur."[89]

Some of the acceptance of tractorization within the USDA, as well as on the part of farm journals, rested upon the sincere belief that the tractor would materially benefit the small producer by enabling him to compete with larger farmers. Under the Agricultural Adjustment Act's restrictions, grain farmers who utilized tractors could sell all their produce, while horse farmers needed a portion of their grain to feed draft animals.[90] While the Department of Agriculture pressed for increased farm production, its officials did not believe such an emphasis would lead inevitably to huge, corporate farms. "We definitely don't want our farms to become soulless factories," stated Secretary Clinton P. Anderson immediately after the Second World War, for "[t]o allow our machines and our impetus toward bigness to bring that about would be to take the American spirit out of farming."[91] Mechanization and family farming were inextricably linked in the USDA's view, perhaps revealing a misreading of the tractor's long-term implications for agriculture. The Department of Agriculture stood for and promoted at once the family farm, scientific farming, mechanization, and increased levels of productivity in agriculture.[92] Indeed, secretary Ezra Taft Benson strongly denied any USDA bias against small farms. He stated, "My whole endeavor as Secretary has been to promote the Administration's efforts to safeguard the family type farm."[93]

As an article in *Southern Agriculturist* highlighted, tractors did hold certain benefits for small landowners and tenants alike. Frank Rudy, a Tennessee tenant farmer, owned no land but did own a tractor. Rudy personified the successful tenant who used his machine "to increase leisure time and diversify crops." He tilled several hundred acres of cropland with his "mechanized mule"—200 acres of wheat, 250 acres of corn, and some tobacco. But Rudy did not hold exclusively to cash-crop farming. He accepted the gospel of diversified farming and the tenets of the live-at-home program pushed by extension agents and farm journals. Apart from his grain crops, Rudy raised lespedeza, clover, a one-hundred-head herd of Hereford cattle, truck and home-use garden produce, pigs, chickens, and horses. Despite his love for horses, Rudy found the benefits of his tractor too numerous to warrant using draft animals. Horses became more of a leisure activity for Rudy. Some of the tractor's positive points included its economy, its belt power, and its ability to work longer hours than horses when necessity dictated. Rudy noted in particular the utility of the belt pulley with which he sawed wood and ground feed, only two of the numerous chores that he "could not do with any number of teams." "Farming with a tractor is a pleasure," concluded Rudy.[94]

Rudy believed his livelihood, prosperity, and success were tightly bound to his tractor. His rejection of draft animals reflected a widening conviction throughout the South—based largely on the midwestern experience and USDA observations, and reinforced more often than not by the southern agricultural press—that successful farmers hitched their equipment to tractors, not to mules or horses. Frank Rudy disproved the stereotypical view that the southern tenant farmer "is typically a victim of oppression whose day is an unending cycle of backbreaking toil, and whose share of the landlord's bounty is a mere pittance." That view of a South peopled with innumerable Jeeter Lesters and Dorothea Lange photography subjects was, according to the article's writer, "still present in the minds of some Northern sociologists, economists, professors and misguided social workers," who were disappointed when they met men like Frank Rudy.[95] To tractor proponents, the machine heralded a way to remake Tobacco Road into a modernized, prosperous South, finally unhitched from inefficient, backward farming techniques and ready to meet the bright future seated on a tractor, not stumbling over clods behind a one-mule plow. The machine, not the mule, exemplified the modern South.

Agricultural journals reflected to some extent the ambivalence that even the most progressive-minded southerners could feel about mechanization. The *Progressive Farmer* was arguably the leading farm journal published in the South. In 1926, it boasted a circulation of 475,000. The editor estimated that there were five readers per subscription, so that the journal had more than two million readers.[96] Its pages reflect the uncertainties, on one level, of farm mechanization. Its July 29, 1922, cover, for example, highlighted two draft horses with the caption, "The tractor has its place, but so have fine animals like these." On another and broader level, however, the journal demonstrates how quickly the various assumptions about what farming should be led to the ultimate demise of the mule, although the final result would not be seen for decades. The journal consistently argued for greater efficiency, and over time that term became equated with tractors and mechanized agriculture. By the mid-1950s, a regular column appeared entitled "The March of Machinery" which reflected changing assumptions about the centrality of tractors and the demise of mules. One 1956 column, for example, featured six new tractor models.[97]

The Corn Belt scenario replayed itself in the South, but by the 1930s and 1940s, the USDA had become more confident that tractorization was inevitable and, in the South, positive, because of the depressed state of southern agriculture. Throughout the process, however, some ambivalence remained about the actual gains and losses for the South. The agricultural shifts which the Midwest underwent during the 1920s, and which fueled much of the clash between draft-animal and tractor proponents, may seem far removed by time and geography from the Depression-era South. Yet, the conclusions that the department developed in the 1920s, the studies it undertook, and the unsure approach it demonstrated toward the issue as a whole carried over directly into its relationship with southern agriculture as the South struggled with mules and tractors in the coming decades. The surprising fact about southern tractorization may be how little the USDA directly involved itself in the process, often preferring instead to watch and study events as they unfolded rather than attempting specifically to direct them. Even in the cotton South, where complete mechanization had to await the development of specialized equipment and techniques, the machine ultimately won out over the animal.

5
Successful Farming Defined
The Tractor Triumphant

Although mules did not disappear overnight from the South, in retrospect their demise seems exceedingly quick. Few observers predicted just how rapidly machines would completely displace draft animals in the region. Man, machine, and animal had for so long coexisted, each filling a familiar, comfortable, and necessary role, that the disappearance of draft animals seemed at least as unlikely as the disappearance of farming itself. The same sanguine outlook had been expressed before as horses faced competition from bicycles, electric trolleys, automobiles, and tractors. Throughout the late nineteenth and early twentieth centuries, far from disappearing, the horse population grew until it reached its apex sometime between 1910 and 1920. At that point, there were 110 million equines worldwide.[1]

Mules simply appeared too useful in southern agriculture to be supplanted by machines. In 1940 a Mississippi county agent related that tractors were used on all the plantations in his area, but that in some years mules could get into soggy fields sooner than tractors and thus plantations using work stock sometimes "made the best showing."[2] For this and many other reasons, the 1955 census showed that the South still had 1,399,567 farms without tractors.[3] Hints of the coming changes appeared in the early twentieth century, but not until the 1930s did a variety of factors begin to converge that limited the mule's relative utility and competitiveness in southern agriculture. By the post–World War II years, the southern mule's fate was sealed, and a return to an agriculture powered solely by draft animals was both unthinkable and logistically impossible.

The automobile and truck revolution of the 1920s had already demonstrated that massive displacement of draft animals could occur, but that transformation was incomplete. Millions of farmers, many of whom lived in the South, gladly left their buggies in the barn and purchased automobiles for transportation to town during the first decades of the twentieth century, but many of them still relied heavily on draft animals to provide most of the farm power. Cars and trucks displaced many work animals, but tractors and other farm machinery such as the cotton picker led to the ultimate decline of draft animals.[4] While the former often merely kept draft horses and mules off roads and byways by limiting their domain to the cotton field or cornfield, the latter pushed them into the shadowy domain of obsolescence.

Observers saw some potential for change in draft-animal usage due to tractorization, of course, but a common theme was that draft animals would continue to play an important role in the South. One writer for the *Progressive Farmer* predicted that "so long as man continues to work land and grow ordinary farm crops, just so long will he continue to draw the lines over good horses where brain, muscle and nerve can be depended upon to 'carry on' under every condition."[5] Even on the eve of massive tractorization, it was impossible to predict the tractor's full impact on agriculture.[6] Thus, until very late in the process that led to the end of the mule, even the most vocal machine advocates did not foresee a southern world devoid of mules.

For decades, southern agriculture appeared impervious to modernizing influences. In 1920, the *Breeder's Gazette* confidently proclaimed, "So long as rice, corn, and cotton are cropped, so long as railroads are built, so long as coal is mined and lumber cut, the mule will continue to be in keen demand as the all-time champion beast of burden for these purposes. The mule is an essential cog in the wheel of industrial and commercial progress."[7] Reflecting the sanguine perspective that consistently remained incapable of seeing a world without draft animals on at least some farms, the *Southern Agriculturist* commented in 1938 that the increased use of tractors and other farm machines did not "necessarily mean a decrease in the use of mules and horses on the farm," since new uses of the animals were constantly being developed.[8] As late as 1948, ten southern states were home to eighty-three percent of the nation's approximately 2.5 million mules, even though farmers in the Old South states bought tractors at a rate exceeding the national average.[9] In 1947, a writer in no less an authority than the USDA's *Yearbook of Agriculture* declared, even in the face of a seventy-five-year nadir in horse numbers, that "those of us who believe there will always be a place for horses in the rural scene have gone ahead with tests and experiments to breed better horses and to learn more about them." Researchers did not focus solely on riding animals, either. Missouri researchers using mules developed an ergometer to accurately assess the amount and speed of work done by draft animals.[10] Draft animals played a key role in industrialization and continued to serve as an integral component of the economy, so there was perhaps logic in concluding that horses and mules would always coexist with machines.[11]

Even the *Progressive Farmer,* despite its name, noted in 1946 that while mechanization had resulted in many draft animals being unemployed, they were not "unemployable." Indeed, idle work stock was a condition "probably more temporary than the machine enthusiast imagines." The author of this *Progressive Farmer* article explained that after the nation reconverted to a peacetime economy, the number of subsistence and part-time farmers would increase, and horses or mules would be the logical source of power for those farms. He concluded that "[t]here will always be a demand for some work stock."[12] An earlier article, however, was more accurate in its assessment of the possibility of a draft-animal renaissance when it pointed out that once the region's farms mechanized, the South "would not easily slip back to the old hand labor, low-living standard basis."[13]

The shift away from mules was not preceded by a thunderclap, but as older farmers and researchers either retired or converted to the gospel of tractor power, the debate shifted from "tractor or mule" to "what kind of tractor?" And as the momentum shifted, the change sped ever faster toward a fully mechanized southern agriculture as the old world finally collapsed, not so much of its own weight but because its foundations had been eroded into oblivion. Many southerners saw a bright future being erected on the ruins of the old, mainly due to the central place of the machine in the new order. Stated one observer, "Yes, I love horses. Used to raise them. But they are on their way out, the tractor is coming in, and a better day for the South lies ahead."[14]

USDA studies and reports demonstrated that efficiency had improved in American agriculture since the founding of the nation, and especially since the turn of the twentieth century. The most significant impact of power and machinery was the ability for farmers to produce more crops with less labor and in less time. For example, the time necessary to harvest an acre of wheat dropped from forty-seven man-hours in 1830 to about six man-hours around the turn of the century. In some areas, the figure had dropped to about two man-hours in 1930. Corn production also required less labor. In 1855, approximately thirty-four hours were required to plant, cultivate, and harvest one acre of corn. By 1930, that figure had dropped to just under seven man-hours in areas suited to mechanization.[15]

Cotton showed improvements in the same direction, but hardly as dramatic because of the labor-intensive work involved during chopping, hoeing, and picking this crop. Still, between 1841 and 1930 the man-hours needed to produce an acre of cotton dropped by one-half in some areas, from nearly 150 man-hours to approximately 72. In most of the South, however, labor requirements per acre had dropped from about 150 man-hours per acre to just over 102. USDA officials assumed that farmers welcomed any changes that reduced labor requirements. According to USDA surveys, "[f]armers were interested in any new power unit or machine which would make them independent of hired labor and permit the timely performance of farm operations."[16] The USDA investigations reflected a genuine interest in the relative efficiency of draft animals to tractors, but the studies also rested on the assumption that speed and good farming were synonymous, thus offering a quantifiable measure for defining what it meant to be successful. One USDA study listed the "Methods of Measuring Success in Farming." These included farm income, labor income, rate of income on capital, cost of production per unit of measure, net income per mule, and yields per acre. The study particularly stressed yields and cost reduction.[17]

Importantly, this attitude was deeply ingrained in much of American thinking, so the USDA hardly ran counter to the American desire for labor-saving devices. This extended to the agricultural sector, where good farming increasingly came to be equated with the number of acres one could plow, plant, or harvest in a day. And it was undeniable that, in empirical terms, tractors could usually win in head-to-head competitions with mules, especially after manufacturers improved upon the first generation of tractors. Tractors improved; the mule remained the same. And to remain the same was to become obsolete.

At first, tractors made inroads in areas of the South where geography offered large, flat acreages. The Mississippi Delta and sections of Texas, for example, tractorized earlier than the rest of the cotton South. A study concluded in 1933 that compared to areas geographically suited to large-scale equipment and farming methods, the labor requirements for the Cotton Belt as a whole had changed little since the turn of the century.[18] As late as 1939, the ever-vigilant Horse and Mule Association of America observed that in the old cotton states, tractors were not replacing mules to any great extent. As the secretary's report explained, "That is cotton country, small-farm country. It is still the colored family and mule and twenty acres per man." In the Delta and other low, flat areas, however, tractors were "cutting in" on the mule's territory.[19] Statistical evidence supported the assertion that tractors were displacing mules. Nearly one-third of Texas farmers, for example, owned tractors in 1945, an increase of approximately ten percent from 1940. South Carolina and Georgia farmers, on the other hand, lagged far behind the 1945 national average of 30.5 percent. Fewer than six percent of farmers in those states reported tractors in 1945, and in neither state did more than four percent of farmers own tractors in 1940.[20] The rest of the South lagged behind Texas and the Delta in terms of tractorization.

Small tractors were designed to fill the needs of the small, one- or two-mule farmer, and manufacturers worked hard to develop and promote these machines as the small operator's salvation, especially through heavy advertising in farm journals. It was not, therefore, the large tractor alone that spelled doom for the mule, but also the small tractor which could perform a number of tasks on the farm. Yet, in the South even the small, general-purpose tractors were too large for single-family cotton plots. In 1940 the average southern farmworker tilled only about twenty-five acres of crops and eleven acres of pasture. In most cases, neither cropland nor pasture on any given farm was of high quality.[21] Thus, erstwhile Agrarian Herman Clarence Nixon correctly observed in 1938, "The steel mule demands more than forty acres. It demands, moreover, fairly good acres."[22] Even small tractors, then, encouraged some consolidation of fields.

On average, a farmer who plowed one acre of land walked and stumbled over eight miles behind his mule.[23] A tractor advocate described his plowing experience in some detail. "How many acres can a man with a mule plow per day?" he asked. Then he answered his own question in terms designed to remove any nostalgia:

> I remember when I was marking off rows, by sighting between the mule's ears at a stake on the opposite side of the field, I could get over about 7 acres of land in a day. Then if I was throwing 2 furrows back over the marker-furrow, I could cover about 3 1/2 acres. It usually took 4 furrows to break the land between rows, so then I only covered 1 3/4 acres per day. If the balk left in the middle was burst out with a solid sweep, I had to make a total of six trips across the field to get the land bedded up. I made an extra trip to put out the fertilizer. Two more trips to throw 2 furrows back over the fertilizer. Then another trip with an old Dow Law planter to plant the cotton. I'd make four or five cultivations with a bull tongue and heel sweep with two trips per row for each cultivation. The total number of trips I'd make to grow a row of cotton was—gee, I've lost count. If you're a mule cotton farmer you don't need to be told.[24]

After a day of plowing or cultivating, a farmer then had to unhitch his mule, rub him down if he was hot, water and feed him, and put him up for the night. If the farmer used two-mule equipment, hitching and unhitching, feeding and watering, and grooming took even longer. This fact was not lost on draft-animal proponents whose main spokesman, Wayne Dinsmore, presciently observed that despite abundant feed and an adequate supply of reasonably priced, good work animals, a significant, if not the most important, obstacle to continued draft-horse use was the fact that draft animals required too much time and attention when compared to tractors.[25]

Dinsmore understood well human nature, and the physical demands of mule farming certainly weighed heavily on farmers. Others understood this fact as well, including some ex-farmers who stayed in the agricultural universe, perhaps in research or teaching, and thus became the purveyors and architects of a more convenient and less grueling farm life. Harris P. Smith, who went from mule farming in his boyhood days to economic research, answered his own question of why cotton farmers were replacing mules with tractors. "Perhaps many," he said, "feel as I did in my boyhood days on a South Mississippi farm. I was dead tired after I plodded along all day behind a mule pulling a Georgia stock and half-shovel plow. Dirt got into my 'brogans,' jammed my toes, and rubbed my feet raw. I wondered if there wasn't an easier way of making a cotton crop." Smith went on to note the drudgery and wasted time, in his view, involved in feeding, watering, and harnessing mules. Even more telling is his description of "creeping along behind an old mule at a speed that gives you ample time for uncovering plants with your toes between steps," while with a "tractor you can sail up and down your cotton rows at 5 miles per hour."[26] But the changing definition of what constituted hard work moved inexorably onward. Just a year later, an author in the same journal touted the advantages of paying an extra $100 for power steering on a tractor. To him, "power steering means the difference between feeling tired or feeling rested at night. I can operate my power-steering tractor all day, go in at night, relax and feel rested." He could not say the same for his other tractor that had no power steering.[27] Speed and comfort proved unbeatable foes for southern mule farmers and mules. By 1960 at the latest, the question was only what type of tractor one needed; draft animals simply were not an option.[28]

There was no doubt that under most conditions, farmers could get field work done in a more timely fashion with tractors. Studies showed that tractors could do more work in less time performing operations such as breaking, harrowing, and cultivating cotton fields. In the Delta region of Mississippi, for example, preparing and planting a one-acre cotton field required 9.3 man-hours and 20.4 mule hours. Those figures were reduced to 4.9 man-hours and 3.3 tractor hours when using two-bottom plows and four-row planters and cultivators—the sizes of implements easily handled by standard tractors.[29] Tractor dealers used this type of information as a sales pitch by pointing out to farmers that tractors translated animal-grooming time before and after the workday into more available field hours.[30] Crops could be put in the ground by tractor farmers in fewer days than by farmers who used draft animals; more ground could be covered in a given amount of time; and tractors provided more power at the hitch for longer periods, including during the heat of the day and at night, than did small hitches of mules. One of the most compelling reasons offered in favor of tractors was that more efficiency led to less time in the field.[31]

The development of the small tractor—such a key element in the disappearance of the mule from cotton farms—did not occur overnight. Since most early tractor manufacturers aimed at the midwestern market, early tractors tended to be huge machines with a very high weight compared to the power available at the hitch. Because of their size, they were unwieldy and, during the early years of development, undependable. The tractor industry was aware of these problems, however, and increasingly focused attention on solving them. As early as 1920, an article in *Agricultural Engineering* predicted that tractors and implements would eventually be designed as a single unit.[32] A few years later, the same publication included an article which called for the development of a general-purpose tractor as the next logical step in farm equipment evolution. While the industry was providing farmers with more power and larger equipment, the lack of a general-purpose tractor limited how that power could be employed.[33] As a result, even when farmers purchased tractors, the tractors could not always be employed enough to pay for themselves. The *Progressive Farmer* noted this fact in 1940. Even while tractor use was increasing, many farmers did not use tractors for enough tasks on the farm. Thus, the magazine suggested that "manufacturers, agricultural colleges, county agents, and farm papers should all work together to see that every tractor owner uses it as he should, and derives the maximum profits from its use."[34]

International Harvester addressed the problem of the general-purpose tractor in the mid-1920s with the development and introduction of the Farmall, the first tractor and implement line designed around specific cropping demands. Farmall engineers reversed the prevailing trend of designing tractors based on accepted engineering practices, choosing instead to focus on the needs of row-crop farmers. Although hardly trumpeted upon its introduction, the Farmall was a radical departure from contemporary tractor design.[35] Introduced in 1924, the Farmall proved revolutionary for both farming and farm equipment manufacturing and, according to one recent study, "doomed draft animals."[36] Between 1924 and 1932, International Harvester produced more than 134,000 Farmalls.[37] A further refinement of tractor design quickly adopted by all manufacturers, the pneumatic tire, made its debut in 1932. Early experiment-station studies demonstrated that rubber-tired tractors used almost one-fourth less fuel and lasted more than one-third longer than their steel-wheeled counterparts. They also offered a much smoother ride to the operator and reduced breakdown-causing vibrations to the tractor itself. By 1939, three-quarters of tractors were leaving the factory with rubber tires.[38]

Competitors quickly moved to produce their own versions of the Farmall, and competition led to refinements such as power takeoffs, hydraulic lifts, integrally mounted attachments, and fast-hitch systems being included on small tractors. John Deere, for example, targeted small farmers during and after the 1930s with models such as the B and H. Almost overnight, the farmer's ability to cultivate eight acres per day with single-row equipment jumped to twenty to thirty acres per day.[39] John Deere produced 306,000 of the B between 1935 and 1952 because of its popularity. Deere advertisements claimed that the company was responding to small farmers who called for tractors suited to their size operations. The sales figures and longevity of the B tend to support this, although some larger farmers also purchased smaller tractors to fill specific needs on their acreages.[40]

The original Farmall met the needs of many farmers, but millions more did not need a tractor even that large. Manufacturers discovered a large untapped market of more than four million small farmers during the late 1930s. Tractor makers surmised that these farmers still used horses and mules because they had to do so. "Four to five million farmers, in the brackets where competition is keenest and cost savings count most, evidently don't use tractors because the industry has not given them an economical power unit," stated *Implement and Tractor.* Tractor manufacturer Allis-Chalmers defined economical power in this case as a small tractor unit for four-horse farms that cost $500 or less. Allis-Chalmers targeted this market with their B, a 2,100-pound, 16-belt horsepower tractor with rubber tires which listed at $495. Two-row cultivation was possible with the B, but it was meant mainly for one-row cultivation of corn, cotton, and potatoes.[41] As such, the B and tractors of its ilk aimed directly at the farmer most likely to employ draft animals.

Despite rapid innovations in the tractor industry, the horse and mule influenced tractor design and marketing. Tractors had to adapt to crop spacings of twenty-two to forty-eight inches, but the most common crop-row spacing of forty-two inches allegedly derived from the mule's anatomy.[42] Even after World War II, many advertisements gauged the power of a tractor not by touting its horsepower per se, but by comparing its capabilities to a two- or three-team hitch of animals.[43] All farmers understood how much work a team could do in a day, but not all could translate tractor horsepower into acres plowed or planted in a day. For a time, horse and mule terminology dictated the language of discussion among researchers, tractor manufacturers, and farmers.

After World War II ended, the emphasis on small tractors continued along with the development of larger, more refined tractors. Versatility improved in both small and large tractors. In 1939, for example, International Harvester introduced the Farmall A model, which produced just over sixteen drawbar horsepower, and the Farmall M, which produced approximately thirty-four drawbar horsepower.[44] The difference in raw power available to the farmer was staggering. In 1930, the approximately 5,200,000 mature mules on the nation's farms provided just over 4,100,000 horsepower. In the same year, approximately 900,000 tractors provided over 22,000,000 horsepower.[45]

Significantly, tractors increasingly had their own lines of equipment that mated with the tractor itself. Farmers did not, therefore, buy only a tractor, but often purchased attachments designed for that particular make and model, as Howard Perdue noted of his McCrae, Georgia, tractor dealership's sales.[46] This had been true to an extent even when tractors only pulled implements in a fashion similar to draft animals, since horse-drawn implements seldom could withstand the beating they received hitched to a tractor. Further, horse-drawn implements limited the type of work tractors could do. As general-purpose tractors emerged, implement engineers played a key role in expanding the tractor's range of work by providing suitable cultivators, mowers, saws, grinders, blowers, and a host of other attachments.[47] This matched tractor-implement "system" proved to be the undoing of the mule in the South because it offered power, convenience, and flexibility. In short, new tractor and implement designs permitted more power to be employed more conveniently and, in purely quantitative terms, more efficiently in carrying out a designated task.

Henry Ford and Harry Ferguson partnered to produce a revolutionary tractor design employing the now universal three-point hitch system. In advertising information, the manufacturer crowed that its tractor "was designed to eliminate the horse. Not to supplement, but to eliminate him because he is a waster of land and time, the primary wealth of the farmer." The booklet explained that experience had proven "that the horse cannot be eliminated by substituting a good machine in front of the same old implements." Thus, the Ford tractor with the Ferguson hitch system provided a convenient means of fully integrating tractor and implement and increasing the tractor's pulling power.[48] Although other manufacturers may have been less pointed in their plans for the horse and mule, they did work toward eliminating the animals on American farms. On some large plantations, tractor manufacturers went so far

Fig. 7. Small general-purpose tractors such as this International Harvester Farmall A played a key role in bringing mule-powered agriculture to an end. (South Carolina Cooperative Extension Photographs, Special Collections, Clemson University Libraries)

as to ship tractors and equipment free of charge for a trial period, hoping to sell the demonstration units should the farmer decide to tractorize.[49] In the end, industrial ideals triumphed on the American farm, and many farmers embraced the changes the tractor wrought upon the southern landscape and on their own acreage.

The postwar years saw more small tractors, both with and without attachments, on the market. This continued the late-1930s trend, and manufacturers regarded small tractors as a logical progression. At the end of the Depression decade, one industry source viewed small tractors as "at least inevitable" since the industry had only recently begun to fill the needs of the two- to four-horse farmer. The plethora of small tractors was expected to "exert far-reaching influences on the entire tractor perspective of the future."[50] The industry was not as concerned with farm size as it was with expanding its markets. Millions of small farms without horses or mules suited the tractor manufacturers perhaps better than a few large farms. Even so, in the 1940s southern farmers were much more likely to plow with mules than with a tractor, and in much of the region, mechanized farming seemed distant indeed. In Edgefield County, South Carolina, the extension service found that "life on most farms still centers around the cotton crop. Farms usually include one or more mules for workstock; one or two milk cows; a small flock of chickens; one or two hogs for pork; a small barn used mainly for shelter rather than feed storage; and a small garden."[51]

Researchers and journal editors exhibited a keen interest in small farms in the post–World War II era. Most wondered whether or not the small one- or two-mule farmer could make the transition to mechanized agriculture. The *Progressive Farmer* noted in 1946 that it was "keenly interested in this whole subject of more power for small farms," and so planned numerous articles connected to it.[52] Wrapped within the issue was the understanding that the definition of the term "small farm" had to change. In short, tractors demanded more acres. Tractors could hardly be used to break a few acres of cotton or tobacco, explained the journal, but "that is not what we want. What we want is to have him [the small farmer] cultivate a considerable acreage of pasture, hay, and feed crops. . . . Tractors make it possible to get away from the old-style, little-acreage, one-mule, all cotton, 'Man With the Hoe' type farms."[53] The *Progressive Farmer*'s treatment of the subject climaxed in 1948 with an eight-part series on mechanized cotton farming by one of Alabama's leading farmers, Price McLemore. McLemore warned that farmers "cannot adopt halfway measures of mule and tractor farming and expect to attain maximum efficiency and results. Having made up his mind to practice mechanized farming, he should forget all his mule farming."[54]

To many, the machine represented salvation for southern farmers. The National Association of Farm Equipment Manufacturers and agricultural periodicals, much more so than governmental agencies, hammered upon the labor-saving and timesaving potential of the tractor and other equipment because they, like Wayne Dinsmore, understood that most farmers would welcome fewer hours in the field and less time and labor in the barn. Farmers, of course, could choose to continue farming with draft animals, but only by sacrificing cash profits as well as time. Arnold P. Yerkes, the editor of *Tractor Farming*, pointed out that a farmer was free to select "any type or size of equipment that may suit his fancy, or even use hand methods if he so desires. Of course he can do so if he is not interested in the returns he will receive for his labor. But if he expects to receive a net income equal to that of the more efficient farmers he doesn't

Fig. 8. Although poorly funded, African American extension agents worked with African American farmers to modernize their farms. Note the power takeoff shaft that transferred power from the rear of the tractor to the implement. (South Carolina Cooperative Extension Photographs, Special Collections, Clemson University Libraries)

have much choice in selecting his equipment—his competitors practically determine what kind of equipment he must use if he is to compete with them on anything like equal terms."[55]

Progress meant higher yields, more efficiency, less backbreaking labor, higher profitability, and a higher standard of living. Comparing regional differences, farm equipment manufacturers found that on average, Illinois farmworkers used $600 worth of farm equipment, while North Carolina's farmworkers used only $115 worth. "The result is just what you would expect," the report observed. Because he used about 8 horsepower, compared to 1.5 in North Carolina, the Illinois worker was able to farm fifty-three acres. The North Carolinian cultivated but twelve acres. In Illinois, therefore, the income was "correspondingly larger." Power equaled prosperity in this equation, because the "amount of power used, which always runs in the same ratio as the equipment, is an infallible measure of farm progress and prosperity." Full of pithy sayings, the report crows, "Farming should be more head work—less hand work."[56]

Tractor manufacturers stressed this theme in advertisements that filled southern farm journals. In a typical declaration, this one promoting the Allis-Chalmers firm, two pictures contrast with one another. A farmer plowing with four mules was "Existing," while a farmer riding a tractor was "LIVING!" The text reads, "Today—you and your family can step into another world. From the gray, drab existence of mule farming, you can emerge into the sunshine of Better Living with Allis-Chalmers power." The advertisement then asks the farmer to decide between a one-crop income with mules or an improved standard of living made possible by Allis-Chalmers products. Another advertisement uses a Tennessee farm family to point out that as a consequence of shifting to tractor farming their "Yields doubled, tripled. Debt disappeared. Livestock flourished. They built a new brick home with electricity and running water."[57] Other manufacturers, from International Harvester to Oliver, echoed similar themes in their own advertisements.[58]

Advertisers and implement companies found hosts of farmers willing to listen to their pitches. Many returning veterans agreed that more power allowed better farming. Exposed to machines of all types during the war, ranging from tanks to the ubiquitous Jeep, many veterans demanded tractors upon their return to farm life. One mule man was quoted as saying, "Them boys wasn't satisfied to come home and follow a jarhead."[59] Veterans could receive preferential treatment in acquiring farm equipment, and many applied for that status in the latter stages of the war, when it was extremely difficult to purchase new farm equipment because of war production demands. In July 1945, 751 certificates to that effect were issued. In August, more than 4,200 farmers received certificates, and February 1946 saw the peak of this program when nearly 61,000 certificates were issued. By far, tractors were the most desired item. Of the 60,917 February certificates, fully 16,265 were for wheel tractors.[60] Many veterans who planned on farming apparently wanted to use tractors, not horses or mules, to work their fields. If they returned to a family operation, their opinion probably proved crucial in convincing the head of the household or farm to sell the mules and replace them with a tractor. And their children would certainly accept nothing less than a fully mechanized farm. Advertisers pointed out that tractors might well keep children on the farm.[61]

Although the USDA produced numbers of publications related to mechanization, the thrust of its studies revealed little about the human element in the mechanization equation. Approaching the subject from a different perspective, Harald A. Pedersen conducted a revealing study of the Yazoo Mississippi Delta region in the postwar era. Pedersen examined mechanization and its impact on farm labor from a sociological perspective. As such, his study focused on "the human element in the progress of mechanization, that is, the impact of the attitudes of farm people and the values of rural society in the Cotton Belt upon technological change."[62] One of the author's most significant contributions stems from the fact that he questioned individuals from all levels of plantation society. He interviewed tenants and laborers, as well as owners and managers. Pedersen examined the plantation because plantations led smaller farms in mechanization and because plantation mechanization led to a greater manpower reduction than did small-farm mechanization. Since the plantation dominated the Cotton Belt, Pedersen's findings contain valuable insights into mechanization throughout the South, although small family farms also had to consider many of the same factors when addressing tractorization.

A complex set of variables determined when and to what degree a plantation owner mechanized. Plantations did not wholly mechanize at once, but adopted machines and methods piecemeal over many years as they sought to balance new technologies with traditional methods and familiar, long-standing social relationships. While all of the thirty-five plantations examined had purchased tractors, each one also retained mules. Only forty percent of planters in the sample believed tractors to be the most economical source of power on their farms. On the other hand, more than fifty percent felt that a combination of mules and tractors provided the most efficient operation of the plantation. Significantly, planters felt that only a fraction of field work—that is, cultivation—could be performed better by mules. More than eighty percent believed tractors performed every other field job, excluding chopping and picking, better than mules. Chopping and picking required human labor, and planters found it more difficult to substitute tractors for humans than for mules. Further, the more a plantation depended upon tractors, the more the operator considered tractor power to be the most economical form of power available. Experience with tractors bred confidence in them, and overall, the planter's opinion of the mule's utility was low.[63]

Even so, several factors flowed against the increasingly powerful currents of tractorization. Many farmers remained skeptical of weed-control methods such as flame weeders and herbicides which were linked to the larger move toward mechanization. Experiments with new methods and materials many times took place on marginal lands and often failed. Farmers rarely experimented on their own land because they considered experimentation to be the responsibility of the universities and governmental agencies.

Nonmechanized farmers also cited costs as a major reason they did not purchase expensive machinery, despite being jeered at by more "progressive" neighbors for owning so many mules. These farmers normally explained that they could still pay their bills, unlike many of "the flame-cultivator and cotton-picker boys." The costs of mechanizing were daunting. Around 1950, tractors and four-row equipment cost $3,000; one-row cotton pickers with the necessary tractor cost $8,000.[64] Not only was horse-drawn machinery cheaper than tractors and tractor equipment; most horse-drawn machinery was paid for. During the early 1940s, it was not unusual for horse-drawn equipment to range from twenty to thirty years old, an average of one hundred to two hundred percent older than tractor equipment of the same general type.[65]

Mechanized farms also sacrificed some quality for the gain in timeliness. Experiments at the Delta Experiment Station in Mississippi involving various types of equipment from a half-row to four-row units demonstrated that cotton land farmed with the smaller equipment produced consistently higher yields per acre than land farmed with four-row equipment. As a result, many planters preferred to use the smaller unit of equipment, which represented "a very small investment as compared to the larger motorized equipment." Operators who used mules could also grow their own feed and, in some cases, produce their own work stock. During times of low cotton prices, these planters avoided large payments for fuel, oil, and repairs to tractors as well.[66] As late as the early 1940s, then, for millions of farmers it still made economic and practical sense to farm with mules.

Utility and cost were not the only factors hindering mechanization. Planters faced a quandary because complete mechanization was costly but piecemeal mechanization had troubles of its own. Tractors demanded a complete overhaul of plantation operations. Cotton could not easily be farmed with both mules and tractors. "Once you start with mules," explained one operator, "it's mules all the way."[67] Further, old cotton gins could not handle machine-picked cotton. Modernizing a plantation's gin to handle machine-picked cotton required a minimum of $60,000. In addition, tractors required new skills, and some planters complained that good tractor drivers were difficult to find and even harder to keep since other planters were not above luring them away with a higher hourly or daily wage. Planters also cited factors ranging from unsuitable land and small fields to the need to retain labor on the plantation for chopping which remained a hand operation at this time. Thus, during the early 1950s, planters generally mechanized "only as much as is necessary to fill the gap between labor demand and labor supply."[68]

Located in the Mississippi Delta, the Delta and Pine Land Company plantation exemplifies the piecemeal approach that even the largest cotton producers took toward replacing men and mules with machinery. Although the Delta and Pine Land holdings encompassed sixty square miles of Delta land, on which lived approximately 1,000 black tenant families or 3,300 laborers, and 1,000 or so mules, it differed from other Delta plantations only in scale.[69] Between 1935 and 1944 the total amount of Delta and Pine Land holdings planted to cotton averaged nearly fourteen thousand acres. The land yielded nearly 630 pounds of cotton per acre for an average annual yield of 17,500 bales.[70] Under Oscar Johnston, who directed operations from 1927 to 1950, the plantation flirted with mechanization during the early 1930s, but after 1933 returned to mules as the chief source of draft power and continued to farm almost exclusively with mules until the end of Johnston's tenure, in spite of the fact that Johnston hired an assistant in 1931, Minor Gray, to oversee mechanization of the plantation.[71] Delta and Pine Land's experience highlights many of the issues involved in shifting to tractors in the cotton South.

An attorney, Johnston ran the company's lands with an eye on profitability and efficiency. His correspondence indicates that he was a perfectionist both in business and personal dealings. After taking control of the plantation, he standardized tenant contracts. After 1927, all tenants worked on a "half" basis. Workers provided only their labor, while Delta and Pine Land Company supplied seed, mules, tools, housing, and credit, along with other tenant needs. The company also provided one-half of the cost of fertilizer and insect control.[72] In late 1931, he began experimenting in earnest with tractors on a small scale, but planned "to go rather extensively into tractor farming" if tractors proved to be superior to mules.[73] Toward those ends, Johnston purchased six new Farmall tractors equipped with optional power lift attachments, four-row planters, four-row cultivators, and disc harrows worth nearly $8,000.[74]

The shift to tractors entailed more than just the purchase of equipment. For example, on one 582-acre tract designated to be worked by tractors in 1932, stumps had to be removed, bridges strengthened, and arrangements made for fueling and servicing the new equipment. Johnston purchased a 450-gallon fuel tank to fit a new Model AA Ford truck which had dual rear wheels and an enclosed cab. Two thirty-gallon side tanks for grease completed the service truck. In addition, the truck boasted a pump with enough fuel hose to pull alongside the tractor while in the field to fill and grease it. Johnston wanted to prevent delays and wear on the tractors by avoiding lengthy trips to a central fueling point during the day. Moreover, Johnston was especially interested in having a means of accurately measuring the amount of fuel each tractor burned.[75] Johnston also had a ten-thousand-gallon fuel tank installed on the plantation to facilitate the handling of tractor fuel.[76]

Johnston could not have been pleased with the results of his foray into tractorization, although looking forward to upcoming years in 1931 he believed that the move to tractor farming would be "quite probable."[77] Tractors proved unsatisfactory, and their use diminished rapidly on the plantation. In 1932 the plantation grew 2,901 acres of tractor cotton. These lands yielded 319 pounds of cotton per acre, far below the average yield for the rest of the plantation. The next year 1,626 acres were worked by tractors, but all of these acres were plowed up, presumably because the expected yield failed to justify continued cultivation and harvest costs.[78] Despite his prediction of 1931, by 1933 he wrote that he questioned "the wisdom of increasing or even continuing 'mechanical farming' on the scale now operated" by the plantation.[79] Johnston had been correct in his earlier prediction that tractorization was part of the future, but the Delta and Pine Land Company would for the most part be a mule and tenant operation until the late 1940s and early 1950s.

In 1943 Johnston reported that his mule-purchase program had "fully justified" his original hope and that it was "quite profitable." He continued the project during 1944, but in 1945 the tide turned. In the labor-short postwar environment, Johnston reported that he had sold "quite a few" mules during 1945 and was substituting tractors for work stock. The plantation's pastures did not go unused, however, for as the mule population decreased, the number of cattle increased. In 1949, 844 Hereford cattle lived on the plantation. Johnston hoped eventually to build a herd of approximately 3,000 breeder cows and market between 2,400 and 2,800 head of cattle annually.[80] Similar changes were occurring across much of the South at the close of the 1940s as cattle replaced mules and horses in pastures.[81] Cropland used exclusively for pasture increased by fifty percent during the decade following 1944, a rate that far exceeded the national average of thirty-three percent.[82]

In 1950, the Delta and Pine Land Company, long committed to mule farming, purchased "considerable farm machinery," including tractors, combines, four-row planters, cultivators, middle busters, and disc harrows to "offset the diminishing supply of manpower." The new equipment replaced nearly 150 mules, which approximated the annual replacement rate on the plantation. The increased use of mechanized equipment resulted in a larger outlay for fuel. In 1934, the Delta and Pine Land Company spent $3,514.82 on tractor fuel. By 1951, that figure had risen to $54,293.74, a total considerably higher than the nearly $39,000 value ascribed to the plantation's remaining work stock. By contrast, the plantation owned $167,852.50 worth of horses and mules in 1938. Although the Delta and Pine Land Company plantation's outlay for fuel dwarfed that of most southern plantations and farms, all farms that mechanized became dependent on fossil fuels which came from off the farm sources. By 1950, the number of work animals on the plantation had been halved from its average during the 1930s. Attrition of labor and mules forced the plantation closer to full mechanization, but Johnston noted that he needed to keep two hundred or three hundred mules on hand "for work for which mechanical equipment is not adapted or suitable in its present state of development." Johnston would retire at the end of the 1950 fiscal year due to poor health. Thus, the end of his tenure coincided with the closing of the mule era on the plantation.[83]

When tractors began to be used in the South, agricultural workers and planters experimented with various means of increasing the number of duties the tractor could perform in the field so that mules would not have to be kept on hand for odd jobs. Plowing and planting quickly became part of the tractor's repertoire, but thinning, cultivating, and picking took longer to become mechanized operations. Cultivation of the cotton crop was added to the tractor's duties next, but thinning and weeding the cotton crop took considerable trial and error.[84] Thinning the cotton plants with a machine was nearly as difficult as picking it with a mechanical device. Near the end of the World War II decade, thinning cotton plants was still largely a hand operation and picking remained almost entirely done by hand.[85] Experimenters came up with methods to cultivate without hand labor ranging from check-row planting to flame thinning. Oscar Johnston began experimenting with flamethrower weeders in the mid-1940s, with an eye toward mechanizing cotton production as much as possible in the face of a worsening labor situation on his company's lands.[86] One must always keep in mind, however, that planters were seldom satisfied with the labor situation. As historian James Cobb has observed, "when planters talked of a scarcity of labor, they often meant a scarcity of labor willing to accept whatever terms the landlord offered."[87]

Again, for all its trumpeting the benefits of mechanization, the USDA was hardly prepared to offer well-researched, proven leadership toward mechanization to southern farmers who faced increasing labor shortages, rising feed prices, and growing foreign competition. E. D. White, an assistant to the secretary of agriculture, recommended that his agency do more to help ease the transition taking place in southern agriculture, but noted that the department was unprepared to do so because it did not understand the peculiar southern situation fully. He recommended a major research effort of four or five years' duration and funded at the level of $1,000,000 per year. White recalled the dif-

Fig. 9. Picking was one of the last steps in cotton production to be mechanized. Until that occurred, bringing in the cotton crop required large numbers of laborers or hands. (Courtesy of Mississippi Department of Archives and History)

ficulties faced by midwestern and western farmers when they shifted to tractors, but explained that in those areas, the farms had ranged from a quarter section to a section in size and the farmers had focused on one crop. He then contrasted that situation with the postwar South, which was "shifting from mules to tractors on small complicated row crop farms. This means abandoning a mule and a double shovel, or two mules and a cultivator, and starting out with a row crop tractor to plant and cultivate perhaps three or four widely varying types of crops." White concluded that "the increased power available calls for a greatly enlarged operation and a new system of timing must be worked out. The complexities arising are numerous and difficult. *Before proper leadership can be given, effective research must be done which will permit the Extension Service to intelligently guide this movement.*"[88]

After the end of the Second World War, the mechanization that the USDA had advocated in a broad form burst upon the South and an unprepared agricultural bureaucracy. During the postwar years, planters—many of whom, like Oscar Johnston, had delayed mechanization either because they understood the process of raising cotton with human labor and mules, or because the machinery had not been perfected to their liking—faced increasing labor and mule shortages. Yet tractors still did surprisingly little of the farm work in the South. In 1939, the amount of land broken with tractor-drawn implements in the old cotton states ranged from seven percent in Georgia to fourteen percent in South Carolina to seventeen percent in Mississippi. In Texas and Oklahoma, states having more in common with the large midwestern farmlands than with traditional southern agriculture, the amount of land plowed with tractors was much larger, around seventy percent in both states. Throughout large areas of the South, in contrast, draft animals broke some ninety percent of the farmland.[89]

The war limited mechanization to such an extent that the percentages changed little until the final mechanization push of the late 1940s and early 1950s. As of 1953, one study concluded that work-stock reductions had been "relatively small" in the Delta, the Southeast, and Appalachian states.[90] Even so, from 1942 to 1952, the number of horses and mules on farms declined by nearly 7.5 million. The decline in horse and mule numbers during that ten-year span exceeded any previous decade's decline. During the same decade, the total number of tractors, including garden-type machines, increased by 2.3 million, while the number of young draft animals born as replacements for aging animals fell year after year. In 1948, the colt crop was only ten percent of what it had been thirty years earlier.[91]

Without enough replacement animals, farmers increasingly had to turn to tractors. This reduced the demand for draft animals and further discouraged breeding. As a result, a downward-spiraling cycle developed, which led to declining numbers of mature draft animals in use on farms and a drop in the number of potential replacements. Importantly, the nationwide trend deeply affected the South since mule replacements were not grown for the most part within the region and southern farmers had as a whole refused to raise their own replacement work stock. In short, the South's dependency for its mules upon the very areas that mechanized first virtually guaranteed that the mule supply would decline as tractors replaced horses in the traditional mule-breeding states. Dependent in large measure upon outsiders for mules, the keystone of its traditional agricultural system, the South would have found it difficult, if not nearly impossible, to maintain a mule-powered agriculture even if all its farmers had desired to do so. World War II briefly heightened the need for draft animals and prompted articles on mule production, but with the end of the war, the last flickers of interest in mule raising winked out.[92]

As the draft-animal supply dried up, so did the pool of individuals who could properly break or train animals for farmwork. Mules were noted for their intelligence but certainly had no innate desire for field work. Knowledge about the intricacies of handling mules and mule-drawn equipment that was passed down through generations of farmers quickly faded in the face of tractorization which demanded its own, but very different, set of skills. A Virginia farm equipment dealer observed in the late 1940s that there simply was no one left to break young horses and mules.[93] While he may have overstated the matter, his point came closer to the truth with the passing of each day.

As the mule supply dwindled, more and more southern farmers, committed to mules or not, found it increasingly difficult to foresee a dependable mule supply in the future. Tractors, on the other hand, were touted as the future of farming, especially in journals such as the *Progressive Farmer.* Eventually the supply of draft animals declined to the point that many farmers had to choose to mechanize or stop farming. Undersecretary of agriculture C. J. McCormick concluded in early 1951 that, for the majority of farmers, "it is now a case of machinery or nothing."[94] While machines did not yet dominate in the South, even in areas still dependent upon draft animals the moment was not far off when "machinery or nothing" would be the case. Even the indomitable Wayne Dinsmore of the Horse and Mule Association of America admitted defeat in the postwar climate of rapid mechanization. The association shifted its focus to pleasure horses in 1947, "because that seemed to be the only field in which we could really accomplish something." Dinsmore noted further that the "work horse and the mule . . . are rapidly going out of the picture. They should not; nevertheless under existing high prices of feed, high prices of labor, and the impossibility of getting men to drive teams, or what seems to be an apparent impossibility, they are declining in use." Dinsmore believed that work animals would always be used, but because the association could not influence prevailing trends, it was turning its attention elsewhere and dropping the work-animal phase of its program.[95] With that, draft animals lost one of their most consistent and committed supporters.

Even in areas suitable to mechanization, the shift to machines was sometimes constrained by local labor conditions. A study of plantation organization and management found that even in areas suited to wholesale tractorization, "the large amounts and uneven distribution of hand labor on cotton have retarded the fuller use of such equipment [and] the extent to which this development can proceed is conditioned by the amount of transient or other labor available for hoeing and picking, and particularly the latter." In lay terms, this meant that cotton farmers needed hand labor, in short, to weed and pick cotton but not to plant. Nor was the need for labor uniform throughout the crop year.

In the Yazoo Mississippi Delta, for example, the cotton crop needed three hoeings during May, June, and early July. Then the cotton would be harvested from late August through November. During other times, the need for labor was relatively light.[96] The cyclical need for hand labor provided some weight against wholesale mechanization. As long as tractors could be used only in limited ways on cotton farms, most farmers would continue to use mules to plow and cultivate, since the labor was on hand anyway. In addition, since the costs related to draft animals were almost entirely fixed—that is, horses and mules had to be fed whether they worked or not—it made economic sense to employ mules as much as possible if they were on hand.[97] Unlike tenants and mules, tractors only "ate" when they worked.

In a sense, the changes prompted by the tractor only prepared the region for deeper paroxysms triggered by the advent of a viable, mechanical cotton picker. If the tractor seemed to herald the last days of the mule, the cotton picker destroyed the vestiges of the uneasy truce that had been hammered out between the mule and the tractor. Indeed, for all of its revolutionary implications for southern agriculture, the tractor simply nicked the bulwarks of traditional farming techniques and the rural social structure. As long as cotton picking was a hand operation, planters needed a large labor supply and worked to keep labor available. Thus, the cotton picker proved to be the truly revolutionary advance that spelled the end of the mule in the South. Over the centuries, cotton growing had become more efficient, but until the mid-twentieth century it had remained a labor-intensive enterprise. The tractor changed that to a degree, but as late as the 1950s, many planters preferred chopping, or thinning, the cotton plants and picking to be done by hand.

Mechanization of the cotton harvest held numerous problems. Machines could not distinguish between healthy cotton plants, weak cotton plants, or weeds. Mechanical pickers degraded the quality of the cotton and picked trash along with the cotton boll. Solutions to the many problems involved "engineers, chemists and fertilizer specialists, plant breeders, entomologists, agronomists, and other scientists." Machine-picked cotton had to be bred to have bolls high off the ground which opened more or less at the same time. Defoliants also had to be effective, as well as the means to apply them uniformly to the crop. Finally, cotton gins needed improvement to enable them to handle machine-picked cotton.[98] The *Progressive Farmer*, intensely interested in developments along these lines, stated in 1948 that weed control and harvesting remained the two main obstacles to fully mechanized cotton farming. This was at a time when approximately 297,000 acres of cotton were machine picked, out of a total cotton acreage of 23,653,000.[99] By contrast, nearly half of the cotton crop was planted with tractor-drawn machinery. During the late 1940s, the acres of machine-planted cotton actually outstripped acres of machine-planted corn.[100]

Not surprisingly, Oscar Johnston, Delta and Pine Land Company's director, was highly skeptical of experimental cotton pickers which began to appear in the late 1930s. The trash picked with the cotton and the amount of cotton left on the stalk was "abhorrent to a producer of premium-price cotton like Mr. Johnston."[101] Thus, in 1937 a Delta and Pine Land tenant could declare, "Yes, suh, we holds by mules."[102] Hand and mule labor generally produced a higher-quality crop than did machines, especially during the early stages of cotton-picker development when the lint could be stained by the juices from crushed leaves and bolls.[103] One study, for example, found that handpicked cotton brought 2.68 cents more per pound than machine-picked cotton. For machine-picked cotton, this represented a loss of $13.40 per 500-pound bale of cotton.[104]

As the labor situation worsened for landowners, mechanization of the entire cotton-growing process attracted more of their attention. At the fore, pushing for a workable cotton picker was the same Oscar Johnston who had been skeptical of cotton pickers in the 1930s. Johnston, however, became a chief advocate of their development a decade or so later. As president of the National Cotton Council of America, Johnston wrote optimistically in 1947 that he felt it "heartening to note that the efforts of those in the fields of research, manufacturing and education are beginning to bear fruit." Among the reasons he gave for the growing impetus behind full-scale mechanization of cotton production were labor shortages, favorable purchasing power, and the development of equipment tailored to the needs of cotton producers.[105] Similar forces were at work in the southern piedmont, although farms there were normally much smaller than their Delta counterparts. Compared to prewar levels, farm wage rates had quadrupled, machinery and fertilizer costs had increased by fifty percent, and motor fuel had risen by fifteen percent by the dawn of the 1950s. Mechanization, explained one report, "which had been retarded prior to 1940 by low wages, low incomes, and poorly adapted machinery, is now proceeding at a rapid rate. The economy of mechanization was doubtful at prewar price relationships, but as wages rise machinery may be used profitably for more and more operations. Better adapted machinery and higher incomes have also contributed to mechanization."[106]

As the labor supply declined and the technology to replace laborers improved, cotton farmers and planters shifted to mechanized cotton farming in the late 1940s and especially during the 1950s. A national study determined in 1948 that in the cotton-growing South, "many croppers and other classes of tenants and wage laborers left the land for other work."[107] The 1948 study reflected the reality of a rapidly changing South. World War II catalyzed profound changes in the South, with the changes in race relations ranking among the most momentous. The civil rights movement ultimately provided blacks with opportunities and choices other than those offered by tenancy and sharecropping and, in some cases, forced them to leave the land. The nascent civil rights struggle had its roots deep in the southern experience, but urban employment opportunities pulled southern workers off farms, gave them a steady income, and exposed them to new comforts and ideas during and immediately after the war. Racial frictions exacerbated by competition for jobs in southern cities sometimes flared into riots and lynchings, but the tide of change was too powerful to stem. Hundreds of thousands of rural southerners left the land during the war, and many never returned.[108] Planters and other landowners confronted the growing labor problem in many ways. Although some tried to keep laborers on the land, most planters found that mechanization held the key to improving production in the face of fewer workers.[109]

It is difficult to generalize how planters and farmers viewed the shift from mules to tractors, but many had misgivings about the mechanization process. In some cases, farmers expected to lose their first mechanized crop.[110] Even if they did not always have total confidence in the new machines and methods of cotton cultivation, however, they often had no choice during the war and immediately afterward. Many made the transition smoothly from mule to mechanized cotton production, but others found it a daunting prospect to completely remake their farms to suit the new machinery. On piedmont farms, terraces often needed to be broadened to enable cotton pickers to be used in hilly fields. The extension service actually discouraged some farmers from mechanizing their operations by opining that parts of the piedmont South simply did not suit the new machinery.[111]

Larger forces pushed farmers toward mechanization if they delayed too long, but then war demands limited the amount of farm machinery produced. Delta Council officials complained that many farms had more tractor drivers than tractors in 1946. Forced because of the wartime labor shortage to mechanize in order to increase production, Delta cotton planters found themselves without equipment and machines because of the wartime emphasis on war goods. As a result, many southern farmers clamored for machinery after the Second World War.[112] In the face of short supplies of farm machinery nationwide and the belief that they were being neglected, members of several county chapters of the Georgia Farm Bureau Federation called on the House of Representatives and the Senate "to immediately instigate an investigation of the distribution of New Farm Machinery and take the necessary steps to effect the routing of a fair share of the New Farm Machinery into the rural areas of Georgia."[113]

The cotton picker itself, however, was long in coming. Its development went back to the late nineteenth century, but by the late 1940s, technology, demand, and related developments intersected to create an environment conducive to an unprecedented reliance upon mechanical operations. While some sporadic, disjointed work was carried out before 1945, after World War II the scientific community and farm equipment industry worked earnestly to make fully mechanized cotton production a reality.[114] USDA official N. E. Dodd wrote in 1947 that the "development of improved mechanical devices for cultivating, including chopping and hoeing, as well as for harvesting is being carefully studied and the Department is giving all possible assistance in stimulating this work."[115]

Delta and Pine Land Company's Oscar Johnston closely watched the development of the cotton picker at least from the time he took over Delta and Pine Land Company's operations in 1927, but was not convinced of its utility until the war years.[116] By the late 1940s, cotton pickers had been refined to the point that Johnston, although disappointed with the wet weather during the harvest season, believed that if not for the use of mechanical pickers, the "result would have been much worse."[117] The growing use of cotton pickers required less labor and fewer mules. Consequently, Johnston's successful program to purchase young replacement mules for the plantation was one of the first casualties of the plan to fully mechanize Delta and Pine Land's cotton crop. Citing the daily worsening labor shortage and "a prospective shift away from [the] use of work stock to the use of mechanical equipment," Johnston indicated that he did not expect to need replacement work stock in the near future.[118] By 1947, the plantation owned 5 cotton pickers and planned to own 150 by 1955. Johnston also was moving forward on a plan to reorganize approximately forty thousand of Delta and Pine Land's acres, "to adapt their drainage, roads and field pattern to the requirements of mechanized agriculture."[119]

The Mississippi Delta, so well suited to mechanized operations, was not alone in its shift to mechanical picking. A 1952 South Carolina study illustrates the inroads made by the cotton picker in the postwar era. Most of the sixty mechanical cotton pickers in the state were located on large farms in the upper coastal plain. In 1949, they proved disappointing. While owners had planned to harvest approximately 176 acres with each machine, they actually mechanically picked only 40 acres due to wet weather during harvest, poor yields, and an adequate supply of hand labor. Here was a peculiar situation of expensive mechanical cotton pickers sitting idle while fields were handpicked. In the early stages of mechanical cotton picking, landowners sometimes preferred traditional methods to modern techniques. Still, farmers continued "their efforts to reduce the drudgery and labor required in growing cotton by more effective and economical use of farm machinery."[120]

While Johnston could foresee the disappearance of mules from his domain and other large plantations, he did not believe that the animals would vanish from the South as a whole. In his estimation, small farmers would continue to grow cotton, employ a significant number of mules, and compete effectively with larger operators who were using mechanical equipment. By filling the demand for high-quality handpicked cotton, the small producer, although unable to mechanize, could compete with fully mechanized cotton producers. According to Johnston, a farm of seventy to eighty acres with around eight acres of cotton under cultivation "will not be driven out of cotton production by mechanization [because] Cotton is still the crop which gives the greatest return per hour of time of any crop he can raise, even if the only mechanization he enjoys is that of preparing his land for planting with the same tractor he uses in producing other crops." In short, small cotton producers had a place in Johnston's future South. Some of these farmers would find it possible to mechanize; others would continue to use mules to produce cotton "the old-fashioned way."[121] While Johnston's predictions may have proven untrue, they do reflect the prevailing attitude held by many southerners that mules would play a diminished but still important role in the future of southern agriculture.

By 1950, despite Johnston's and others' belief that mules would not disappear, the question for most farmers was no longer whether to shift to tractors, but what kind of tractor to purchase.[122] Farmers who gave up mules either traded them in on tractor equipment or sold them outright. Not all dealers accepted mules on trade-in, but many did.[123] The shift away from draft animals meant that horses and mules were no longer needed, and slaughterhouses moved in to take advantage of the situation. Thus, the rapid displacement of mules and horses led to the destruction of thousands of healthy animals during the 1940s and 1950s. This slaughter inspired a few complaints and calls for investigations into the matter. Thousands of jacks were sold as food for minks and foxes on northern fur farms.[124] A College Station, Texas, agricultural experiment station economist noted in 1954 that "dogs and cats are eating the cotton farmers' mules."[125] Wild horses on the western range and "surplus horses which have been replaced with motor-driven equipment by progressive farmers" were the major sources of animals for the slaughterhouses.[126] Evidently, only a few people protested this development, but it was a supremely ironic moment when draft animals ceased fueling the nation's agriculture and instead became food for the fur and pet industries.

Concerned correspondents to the Department of Agriculture worried about the welfare of the animals, but they also argued that the disappearance of draft animals would weaken the nation's agricultural system. One man warned against the "wanton slaughter of fine, young, serviceable draft horses that one day in an emergency may be sorely needed in our cities and on our farms."[127] The USDA dismissed such complaints as insignificant, since draft animals clearly no longer served a purpose. If they did, the department reasoned that "the slaughter of horses would not have developed into the extensive industry that it has."[128] Eventually, the issue faded as horses and mules disappeared.

By the 1950s, the mule's fate had been sealed. In 1962, the USDA's Statistical Reporting Service stopped enumerating the number of mules on farms because they played such a small role in agricultural production.[129] After this date the remaining mules, now officially nonexistent, worked out their lives in small plots or, if fortunate, retired to pasture and barn lots to spend the rest of their days as wards of the farm where they once played a central role in providing for the farmer and the nation. Occasionally the retired animals might pose for a local newspaper photographer while ritually plowing a garden in the spring, often followed by an aged farmer who refused to accept the triumph of the machine and whose knowledge of the mule South would soon die with him. At other times, the mule owner might be someone intent on preserving the animal as an aspect of traditional southern culture. Richard Weaver, a professor of English at the University of Chicago from 1944 until his death in 1963, always went home to North Carolina after the spring term ended. His garden awaited him in Weaverville, plowed by a mule or horse at his insistence.[130]

When mules disappeared from southern fields, few farmers went into mourning, but some nostalgic treatments of the mule did appear in the popular press.[131] For the most part, however, southerners who commented on the shift regarded the mule's demise as a logical and normal part of progress. "We may be able to slow down the new development in the South by subsidizing the inefficient one-mule cotton grower," opined a writer in *Harper's Magazine,* "but we cannot prevent the shift to mechanized cotton raising, livestock farming, and timber growing."[132] Understandably and correctly, the focus usually centered on the social repercussions of mechanization such as displacement of tenants, changing race relations, and need for cities and industry to absorb millions of people, rather than on the fate of the mules themselves. In a word, most people in the postwar climate viewed the mule South as anachronistic, obsolete, even embarrassing, and were pleased to see it pass, if they considered the issue worthy of thought at all. To these people, mules were symbols of a South that needed to disappear.

6
The Transitional South
Mules, Metaphors, and Modernization

Though the mule has held many meanings inside and outside the South, within the South itself—a region where symbols frequently arouse intense emotion—the mule's symbolic history is complex. Mule use cut across race and class lines during most of its years in the South, and while it did not unify southerners, it provided a touchstone of experience for practically all the region's inhabitants and was one element of a distinctive regional culture.[1] Mules furnished common ground for black and white southerners, especially after the Civil War. The mule, in short, became a fundamental part of the rich and varied cloth of the American South.[2] And if the South "has been a kind of sphinx on the American land," then the mule was often, and has continued to be, just as inscrutable—a sphinx, if you will, with long ears.[3]

The mule was one of several organizing principles around which southern agriculture pivoted for a century. Journalist Ben Robertson, who died in 1943, wrote of his upcountry people, "We are a primary people, we believe in tangible things—in abstract thought, but in tangible things. In faith, in love, in cotton bales, in acres of lands, in mules."[4] Mississippi businessman and author David Cohn, writing of the Mississippi Delta in 1935, noted that cotton and mules played a central role in his world, too. "Mules were everywhere," he stated simply.[5] The *Southern Agriculturist* observed in 1938 that "the mule is as distinctly a natural product of Dixie as cotton, sugarcane, tobacco and rice. There is no other place where the mule fits in the whole scheme of farming quite as well as 'down South.'"[6] For much of the first half of the twentieth century, the mule represented something solid in southern life, something dependable, and something that marked their region as distinctive.

The mule was most importantly the region's chief draft animal and an essential ingredient of making a living in the cotton South. As such, the mule plowed the fields and pulled the cotton and corn planters and, later in the sea-

Fig. 10. Mules not only plowed the fields that grew cotton; they also transported cotton from field to gin. (Courtesy of Mississippi Department of Archives and History)

son, the cultivators. It then hauled the cotton to the gin and to market. In between these duties, the mule performed sundry tasks on the farm and plantation, ranging from hauling firewood to pulling the family wagon to church on Sunday. So important was the mule to many southern farmers, especially during the dark years of the Depression, that loss of the family mule meant the loss of the crop.[7] In turn, a ruined crop could mean hunger, debt, and foreclosure. A Depression-era North Carolina farmer explained that if his mule died, it would leave him "flat in the world."[8] Erskine Caldwell's Jeeter Lester in *Tobacco Road* finds himself in just that predicament, as did many "original Lesters" who populated the South.[9]

Mules pulled plows, wagons, and other implements, but their uses went beyond the purely muscular. Mules were considered sound collateral, for example. At times, particularly during the early stages of tractorization, mule farmers could get loans when tractor farmers could not.[10] If a loan could not be paid, or if a farmer had other debts he could not meet, the sale or forfeiture of the farmer's mule could help offset the liability. In 1936, C. D. Bolton Company, a general merchandise store in Wilkes County, Georgia, accepted a mule from

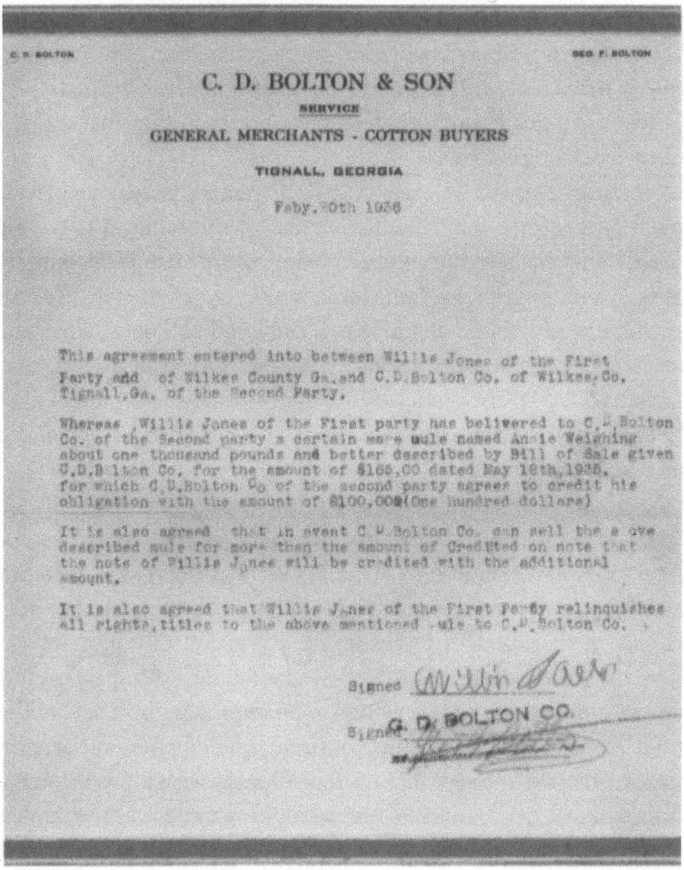

C. D. BOLTON　　　　　　　　　　　　　　　　GEO. F. BOLTON

C. D. BOLTON & SON

SERVICE

GENERAL MERCHANTS - COTTON BUYERS

TIGNALL, GEORGIA

Feby. 20th 1936

This agreement entered into between Willis Jones of the First Party and of Wilkes County Ga. and C.D.Bolton Co. of Wilkes, Co. Tignall, Ga. of the Second Party.

Whereas, Willis Jones of the First party has belivered to C.D.Bolton Co. of the Second party a certain mare mule named Annie Weighing about one thousand pounds and better described by Bill of Sale given C.D.Bolton Co. for the amount of $165.00 dated May 18th, 1935. for which C.D.Bolton Co of the second party agrees to credit his obligation with the amount of $100.00 (One hundred dollars)

It is also agreed that in event C.D.Bolton Co. can sell the above described mule for more than the amount of Credited on note that the note of Willis Jones will be credited with the additional amount.

It is also agreed that Willis Jones of the First Party relinquishes all rights, titles to the above mentioned mule to C.D.Bolton Co.

Signed [signature]

Signed C. D. BOLTON CO. [signature]

Fig. 11. Agreement dated February 20, 1936, transferring ownership of a mule named Annie from Willis Jones to C. D. Bolton Company, Tignall, Georgia. C. D. Bolton Company was a successful mercantile business for much of the twentieth century. (In possession of author)

Willis Jones for debts Jones owed. Jones received $100 credit toward his obligation, but if C. D. Bolton Company sold the mule for more than $100, the extra amount would be applied to Jones's account.[11] Mules also played economic bellwether. High mule prices signaled good times, while low prices meant that the southern economy was sluggish.[12]

Millions of southern farmers relied upon their own strength, that of their family, and that of their mules to wrest a living from the land. Thus, the mule symbolized the earthy side of southern culture. The mule remained, for the most part, a work animal not suitable for riding or pulling a carriage of those in authority. Just as overseers had done in the antebellum South as they supervised slaves, managers on Delta and Pine Land Company's plantations often rode the relatively small number of company horses as they made their rounds to check on black employees.

One old-timer recalled of early-twentieth-century Mississippi, "While the hired hand backed up a pair of mules to the wagon to go to town, the landowner saddled up his best horse or hitched a pair of bays to the buggy. Mules did, however, pull buggies, and very well, if not [in] great style."[13] Hard times could turn the world upside down. When William Faulkner's Judith Sutpen in *Absalom, Absalom!* ventures to town during the Civil War, she wears a made-over dress and rides in "the carriage still but drawn now by a mule, a plow mule, soon the plow mule, and no coachman to drive it either, to put the mule in harness and take it out."[14]

Horses for the most part symbolized authority and wealth in the South, as opposed to mules, which connoted low status. In "Barn Burning," Faulkner describes the Snopeses' landlord as a "linen-clad man on a fine sorrel mare," a stark contrast to the story's ill-nourished mules.[15] Mississippian David L. Cohn recalled that in the 1930s "[n]o encounter was so warmly pleasant as, riding homeward on a horse at first dusk, to meet field hands riding the mules to lot, each politely tipping his cap to the boss man and saying softly, 'Good evenin,' Mistuh Ed.' "[16] This is not to say that southern sharecroppers and tenants never used horses, nor does it mean that southern horses were all well cared for, but horses stood apart from mules in the southern symbolic universe, had done so from the early days of mule use in the South, and continued to do so until the end of the mule era. As one child of the mule South wrote, "Where I come from . . . horses were playthings that few people could afford; mules put grits on the table and bought the baby's shoes."[17]

Perhaps because mules were so crucial to their livelihood, southern farmers reserved a special place in their hearts for them. Ned Cobb, widely known also as Nate Shaw, loved the mules he owned and worked. Cobb's life nearly matched the span of years during which the mule reigned and disappeared in the South during the twentieth century, and he had a lifelong love affair with mules. "O, my mules," he recalled, "just granted me all the pleasure I needed, to see what I had and how they moved." Not surprisingly, tractors held none of his interest. Cobb "was a mule farmin man to the last" who never made a crop without mules. While his son became a tractor farmer, Cobb decided against trying to make the transition to mechanized farming, although he thought through the issues involved. In the end, he concluded that he knew as much about mule farming as anyone, but "when they brought in tractors, that lost me."[18] It was, in his estimation, a profoundly different way of farming. Indeed, he innately understood what more educated observers also saw. "With a tractor," concluded one researcher, "your way of life is changed, you even speak a different language."[19]

The older generation which had been raised during the mule's ascendancy certainly felt differently about farming with mules than did the generation that grew up with the automobile and tractor. Farmers familiar with the intricacies of harnessing and plowing mules complained of the new generation, "Those boys don't know a whiffle-tree from a hame string. They don't know how to talk to a mule. They don't even know how to curry one or put the gear on him. They'll crank up a two-thousand-dollar tractor to do some little old piddling job that could be done better by a twenty-five dollar mule."[20]

To the older generation, as well as to the Vanderbilt University Agrarians, the mule symbolized the strength of southern culture; the region's ability to survive adversity; and a plain, commonsense approach to life and farming. Mules were alive and had to be dealt with as fellow creatures. Joshua Lee grew up on a Georgia farm in the 1920s and 1930s. He understands the older generation's love of mules. "I don't think," Lee notes about his grandfather, "Old Cap'n would have gotten much satisfaction out of petting a six-row tractor rig."[21] This attitude toward draft animals, though it held on longer in the South, was not unique to the region. A 1930 study revealed a connection between the age of farmers in New England and their attachment to their farm horses. Many older farmers, those more than forty-five years of age, kept their horses until the animals died of old age, even after the farmers had purchased an automobile.[22] The same held true with farmers in the South and their mules.[23]

Mules often received special attention. Sam Bowen, a white South Carolina farmer who lived upstate in Anderson County, left a mute testimony to this in his extant farm journals. During the 1920s, Bowen usually owned several horses and mules. He had his draft animals shod at regular intervals and kept close tabs on the costs. In his account books, Bowen normally recorded the name of each mule he had shod, which was usually done in pairs, although sometimes he had more shod at one time. Pearl and Ola, Mex and Annie, Bess and Pomp, and other mules show up time after time. Never, however, does Bowen record the name of a horse. While he may have had fewer horses to keep up with, still it is noteworthy that he felt it important enough to record the names of his mules. Further, his wife appears simply as "wife" in his ledgers when she received cash for purchases. While he doubtless had good reasons for listing animals and people as he did, Bowen must have thought rather highly of his mules.[24]

Mules died from old age, diseases, and accidents, and a particularly ill-natured mule would find himself for sale, but many of them remained with a single family for years, becoming in the process more than work animals to the farmers. Farmers understood that they could not make a crop without the power of the mule, and the walking plow nature of much of southern agriculture demanded closeness between plowman and mule, each depending upon the other for the necessities of life. A North Carolina farmer noted during the Depression that he had raised twenty-six crops with his mare mule. To his recollection, she had never been sick. His mule was a steady, dependable laborer with whom the farmer had spent countless hours in the field and on the road. Each knew the other's idiosyncrasies and routines, and that made work go more smoothly.[25] Ned Cobb kept a favorite mule eighteen years. He said of her, "Kizzie" was as "good a mule as I ever walked behind." His years of farming with her were pleasant for the most part, particularly because "she held up good and I held up good."[26]

When an animal belonged to one owner for years, habits became etched in both the human's and the animal's minds. Mules liked routines, especially related to mealtime. Stories abound about mules that would not plow another step when the dinner bell sounded. "Bout twelve o'clock," Ned Cobb's mules would "ask" to be fed.[27] Mules also developed peculiar work habits to meet the needs of owners. Novelist Harry Crews's family bought a $20 mule from an octogenarian farmer because that was all they could afford. While Pete the mule worked willingly enough, his previous owner had needed to stop every seventy yards to rest for several minutes. Pete continued to follow the same routine in spite of the fact that his new plowman needed no such rest. But the family's hired hand, Mr. Willis, saw no reason to try to change twenty years of the mule's habit.[28] Pete "understood" what his original owner had needed. While some mules may not have worked for a single farmer for decades, thousands of mules stayed on the same farms their entire adult lives, and as the seasons came and went, they became a familiar part of the farm routine.

For all of the camaraderie between man and mule, farmers sometimes envied the mule's station in life, while cursing their own. A common story illustrating this circulated the South in many different versions:

> Over the hill trailed a man behind a mule drawing a Dixie plow. The clodhopper was "broadcasting." "Bill, you are a mule, the son of a jackass, and I am a man made in the image of God. Yet, here we work hitched together, year in and year out. I often wonder if you work for me or I work for you. Sometimes I think this is a partnership between a mule and a

fool. For surely I work harder than you do. Plowing here, we cover the same distance, but you do it on four legs and I on two. So, mathematically speaking, I do twice as much work per leg as you do.

"Soon we'll be preparing for a corn crop. When the corn is harvested I give one-third to the landlord for being kind enough to let me use this corner of God's universe. Another goes to you, and what is left after the weevils and mice have had their share goes to me. But you consume all of yours except a few cobs. I divide mine among seven children, six hens, two ducks and the bank. Bill, darn you, you are getting the best of me; it ain't fair for a mule, the son of a jackass to rob a man, the lord of creation, of his substance. And come to think of it you only help to cultivate the ground. While me and the family is hard at work pulling fodder for you on a hot August day, you are over there in the pasture giving us the hee-haw.

"All fall and part of the winter, the whole family, from granny on down to the baby, pick cotton to help raise money to buy your harness and pay interest on the mortgage on you, and anyway what do you care about mortgages. It doesn't worry you any, not a darn bit. You leave that to me, you ungrateful, ornery cuss."[29]

Southern mules and blacks shared a close identification in the minds of many southerners. From the beginnings of widespread mule use in the South, whites associated blacks and mules. Because planters first adopted mules, slaves and mules remained concentrated in the same hands during the antebellum era. The mule and the slave came to be linked in the popular mind more closely than the slave and ox, or the slave and horse—partly because many observers believed that blacks and mules shared certain attributes. Slaves worked with mules daily, and both were viewed as beasts of burden by their owners and overseers. Both were hybrids of a sort, as far as most whites were concerned—slaves were both person and property, while mules were both horse and ass—each fully the property of the owner who could work, sell, trade, or abuse them as he saw fit.

Hybridization was an important theme in the South, and the irony fairly drips from the fact that a region obsessed with purity of blood selected an animal of mixed heritage. In Faulkner's *Absalom, Absalom!* for example, Charles Bon's tainted blood is perhaps the central driving force of the story. In addition, the reputation of the mixed-blood mule as ornery fit well with the stereotype of "mulatto rebelliousness" as reflected in antislavery novels such as *Uncle Tom's Cabin.* Often, dark-skinned blacks were portrayed as obedient, while mulattoes were, as Dr. Josiah Clark Nott, antebellum physician from Mobile, Alabama, put it, "sometimes prone to revolt in the seething desire to break the chains of slavery." Throughout his writings, Nott explicitly linked mulattoes and mules. Arguing that mulattoes would become sterile over time, he explained the condition as similar to the sterile mule offspring produced by the cross between a horse and a jackass.[30]

Planters believed that both slaves and mules were admirably suited for laboring in the Deep South. Slaveholders consistently maintained that blacks could withstand southern summers more readily than whites. Antebellum planter Joseph Jones, in an address before the Cotton Planters Convention of Georgia, pointed out that "no race but the African can ever stand the burning heat and fatal miasms [*sic*] of the Rice fields, and of the Cotton fields."[31] Mules were similarly branded. Explaining why mules were more common in the Deep South than in other areas of the region, a mid-nineteenth-century commentator observed "that the horse attained a higher degree of excellence in a temperate section, while mules and the darkey were fitted for the South." In the same vein, the *American Farmer* discussed the "greater liability of negroes to winter epidemics," as well as how best to keep slaves healthy through the cold months of the year.[32]

The *Southern Cultivator* advised slave owners in 1853 to treat each servant as a human so as not to make him "understand that he is merely a mule, and possessing only the instincts of brute creation."[33] Yet, slaves were often identified as brutes because of the manner with which they attended animals. Frederick Law Olmsted observed simply "that horses cannot bear the treatment that they always *must* get from negroes."[34] If horses were too refined for the slave to handle, mules were not. Thus, the *Southern Planter* could ask in 1850, "Who, that can remember, has not heard the remark, that mules were the very things for servants?"[35] Harvey Riley, who spent a lifetime working with mules outside the South and served as superintendent of the government corral in Washington, D.C., concluded that mules and slaves were by nature suited to one another because "the mule seems to understand and appreciate the negro; and the negro has a sort of fellow-feeling for the mule. Both are sluggish and stubborn, and yet they get along well together. . . . A negro has not much sympathy for a work-horse, and in a short time will ruin him with abuse, whereas he will share his corn with the mule."[36]

Ohio farmer T. J. Warder noted in the mid-nineteenth century that many people saw a natural connection between mules and blacks. According to Warder, observers were "under the impression that the obstinacy and hardiness and endurance of the one, were naturally adapted and related to the low degree of intelligence and brutality of the other." In this case, Warder defended the mule as a victim of its own "wonderful powers of endurance that enable him to bear unnumbered cruelties which are heaped upon him . . . from contact with an ignorant race, degraded by slavery and its attendant depressing and demoralizing influences." Unlike mules in other countries, Warder continued, those in the United States held an "oppressed and degraded position."[37] Although his argument was convoluted, Warder operated on the assumption that blacks and mules always went hand in hand.

African Americans and mules remained linked to one another in white southern minds after the Civil War and well into the twentieth century. Sharecropping only strengthened the ties between the two. Minor Gray, who worked for the Delta and Pine Land Company in the Mississippi Delta, recalled hearing the tenants "singing in the field and then later on they would tie their mule to the front steps and eat breakfast because their crop was right around the house and they would bring the mules up to the barn at noon and sit under the wagon shed and talk while the mule had his lunch you know and they would go home and plow some more and eat later in the day. It was a real interesting operation."[38] Fellow Delta and Pine Land employee J. W. Fox bluntly observed during early discussions about mechanizing the Delta and Pine Land plantation that "it takes niggers and mules to make cotton."[39]

Similar attitudes echoed across the southern landscape, and they remained deeply rooted across time and place. Racism continued to link mules and blacks, as it had since before the Civil War. "There's just something about a mule," explained one southern observer in the 1930s, "that fits ploughin' a field with a nigger behind him. They're both built for heavy work—that is if you ain't in a hurry."[40] On hearing about a two-row riding cultivator, some farmers scoffed at the idea of providing such a labor-saving device to black tenants, since "they'd go to sleep riding up and down the row!"[41] Popular epithets common in the South—from "stubborn as a mule" to "grinning like a mule eating briers" to "looking like a jackass peering over a whitewashed fence"—could have incorporated racial stereotypes easily.

Less overtly racist and more polished, but embodying some of the same assumptions, works such as Donald Davidson's "Still Rebels, Still Yankees" also coupled blacks and mules in the South. In an essay that Paul Conkin describes as "one of the greatest literary products of the agrarian impulse,"[42] Davidson wrote that rural Georgia "was a well-tilled country, where you were forever seeing the Negro and his mule against the far horizon." Later in the essay, the author describes the view from Cousin Roderick's porch as cotton wagons passed, "heaped high with the white mass of cotton and a Negro or two atop [accompanied by] the jingle of trace chains and the clop of mule hoofs on the almost brand-new state highway, which is so much better for rubber tires than for mule hoofs."[43]

Historian Thomas D. Clark linked mules and blacks. Writing in 1929, Clark noted that "[t]he mule was, and is, just as essential to the cultivation of cotton as the negro was and is now."[44] And when Quentin Compson, in the novel *The Sound and the Fury,* travels home from Harvard during Christmas vacation, he sees a black man on a mule as the train passes through Virginia early one morning. At that moment, he knows that he was back in the South again.[45] Faulkner wrote that the man and the mule waiting for the train to pass had "that quality about them of shabby and timeless patience, of static serenity; that blending of childlike and ready incompetence and paradoxical reliability that tends and protects them it loves out of all reason."[46] Blacks driving mules in a field, blacks riding mules, or blacks in mule-drawn wagons became a firmly rooted element of the southern physical and mental landscape. Faulkner's fictional world describes the reality of the South. David Cohn wrote of the Depression-era Mississippi Delta, "Our roads were alive in autumn with mule-drawn, cotton-laden wagons creaking toward the gins, a Negro driving and often a companion asleep on the piled whiteness who seemed to go—so he was—Elijah-like upon a gossamer cloud."[47] Blacks and mules remained as central to the popular portrait of traditional southern life as cotton itself.

While whites held well-known views of mules, so did African Americans. Slaves and later freedmen remained in close contact with the animals throughout the mule era, and the mule played an important role in African American life and culture.[48] For the most part, slave lore did not focus on mules, but mules were such a common element that their presence was unexceptional. Slave narratives, for example, refer to mules often, but in most cases the mule adds context to the story, rather than being the linchpin of the tale. Recalling her antebellum childhood as a slave, Julia Henderson related a typical story to an interviewer in which her grandmother was plowing, "and the mule go all right, but when it come back to dis end, he would make a dart, and dat would jerk de fer (furrow) crooked, and den she went around to see what make dat dart, and a great big coach-whip snake big as she, jus' rear up and look at her!" She then killed the snake but was whipped for stopping work.[49]

Slave Caroline Ates recalled that she lived about thirty-seven miles from Macon, Georgia, "so that all the crops wuz hauled there in great big wagons with six mules ter the wagon."[50] In a similar vein, Georgian Robert Kimbrough remembered that when he was a slave during the 1840s and 1850s, large numbers of planters would make "pilgrimages" to Savannah, Georgia. He recalled that at times a caravan of one hundred four- and six-mule wagons loaded with cotton bales would trundle toward the coast.[51]

Many African Americans no doubt had a complex relationship with their mules. The animal was a close work companion for most blacks engaged in agriculture, and propinquity can breed affection, especially if the mule driver feels as shackled to his station in life as the mule does to the plow. But the mule also represented dashed hopes in the aftermath of emancipation and unfulfilled promises of freedmen receiving "forty acres and a mule." William T. Sherman's Special Field Order No. 15 may have been the germ of this idea which quickly became a hope for freedmen.[52] Sherman decided to lend freedmen horses and mules, in addition to setting aside land for their use. But freedmen's hopes dissipated in the face of strong countervailing forces which prevented widespread confiscation of southern lands and their subsequent distribution among newly freed blacks. In the end, freedmen found it much easier to acquire draft animals than land.[53]

According to one story told by an ex-slave, Reverend W. B. Allen, whites took advantage of the freedmen's hope of land and a mule. Allen related the tale, perhaps apocryphal, of the Yankee who promised to secure a deed for forty acres and a mule for the cost of $20. Unbeknownst to the illiterate freedman who paid for the deed, it was but a piece of paper that read: "This is to certify that this Negro has been able to secure a piece of paper called a deed to forty acres and a mule, and I hope that he gets both some day."[54] Significantly neither Sherman, the freedmen, nor anyone else called for "forty acres and a horse" or "forty acres and an ox," although both types of animal were still used on farms in the South and both were consistently cheaper than mules.[55]

Many of the works that deal with slaves, tenants, and mules stress the point that mules made such inroads in the South because they withstood abuse well, and this is a point well taken. Likewise, much has been made of the fact that slaves and sharecroppers had little incentive to care properly for their charges.[56] While many mules suffered grievous abuses at the hands of tenants and sharecroppers, many others doubtless received much attention and the care they needed. Ned Cobb cared well for all the mules he owned during his farming career, regardless of his status as sharecropper, tenant, or landowner. Cobb not only made sure his mules were well fed and watered, but before he bridled his mules for the day, he carefully brushed and curried them. On days Cobb plowed with a mule, he cleaned the animal up at the beginning and end of work. Cleanliness was nearly as important as keeping the animals in thrifty condition. "Them animals," recalled Cobb, "my mules or my horses, I considered the next thing to my family."[57] The appearance of a man's mules was important. To many people, "a man who would mistreat his mules would mistreat his family."[58]

Whether or not a farmer cared well for his animals, mules played a key role in his existence. Some scholars have uncovered the importance of mules in their explorations of the black experience through songs and folklore. Herbert Richardson has analyzed a century of black work songs which came from a body of folk music "generally acknowledged as a reflection of the American Black experience." Music played a key role in sustaining blacks through the trials of slavery and freedom. As Richardson says, black workers "used folk songs to maintain group solidarity and articulate their social and economic concerns in a hostile environment. At work, at leisure, and during religious and social events songs occupied an important place in the lives of Black people." Richardson has identified six long-term or recurrent themes in black workers' music, one of which was mules.[59]

Songs about mules reflected a wide spectrum of situations ranging from confronting uncooperative animals to admiring a team of outstanding mules to properly caring for mules. Much of the emphasis on agricultural modernization during the early twentieth century focused on placing larger teams of mules in the field. Switching from a single mule to several could be daunting and difficult, as the song "Mule Blues" points out:

You know what a four mean?
That's a fourspot;
I'm gonna do the best I can,
I haven't worked that many mules,
but I believe I can do it.

Oh, you know I'm workin' so hard, look here,
buddy, you know I'm almost burned out,
Oh, buddy, I just can't hardly catch my wind,
Oh gal, oh gal, wonder why you treat me so mean,
Oh, you know, buddy, this pair o' mules is bout to carry me down.[60]

Recalcitrant mules were a common theme in African American songs, and one must wonder how closely slaves and sharecroppers identified with the plight of an animal that was independent enough to not always want to work, but which was considered expendable to those in authority. A southern black explained to researcher William Ferris that the rule of thumb was once, "Kill a mule, buy another. Kill a nigger, hire another!"[61] Such realities generated a comradeship of sorts between man and animal, and mules must have heard many a heart unburdened in the field, out of the earshot of any other human. The blues song "Trouble" exemplifies such comradeship, although Richardson points out that it is uncertain whether the song is directed at the mule or if the singer's thoughts remain his own:

Trouble, trouble, trouble,
done had trouble all my days,
Trouble, trouble, trouble,
Done had trouble all my days.
Seems, boys, like dose troubles
Gwine carry me to my grave.

Hey, git up dare, mule. Come on, Sally!
Git your head up there, gal. Gee! Haw!
Steady, steady.
Whoa now![62]

Despite an often real affection that southerners held for mules, the animals were hardly excused from punishment. The biblical imperative of sparing the rod and spoiling the child was interpreted often to apply doubly to mules. In story and song, unwilling mules were often persuaded to work by striking them in the head with a sizable object, often a large stick, timber, or singletree. Common to many African American folk songs is a version of the following lines, which allude to one of the most common views of the mule: "Hollered at ol' Beck and she wouldn't gee, / Hit her on de head wid de singletree."[63] Perhaps blacks saw themselves in a similar vein when getting the same treatment meted out to countless "Becks" across the South.

One symbol of progressive agriculture in the black community, the Jesup Agricultural Wagon, or simply the "Jesup wagon," was named after the New York banker who helped fund the enterprise. It was pulled by a pair of mules. During the early twentieth century, filled with items such as a modern cream separator, milk tester, two-horse steel beam plow, and garden tools, the wagon and black county agent visited black farmers who lived in the Tuskegee Institute region of Alabama, urging them to adopt a more diversified and scientific agriculture. The original wagon broke down after fifteen years of service and was replaced with a motorized truck, nicknamed the "Knapp Truck" after Seaman Knapp, the father of the county agent system in the South. Eventually the idea became more widespread.[64] But the mule remained the symbol of poor and downtrodden southerners through much of the twentieth century. So close was the symbolic relationship between blacks and mules that a pair of mules pulled Reverend Martin Luther King Jr.'s funeral wagon. After the service at Ebenezer Baptist Church on Auburn Avenue, King's remains were placed in a sharecropper's cart drawn by two mules, Ada and Agnes, for the four-mile trek to the campus of Morehouse College.[65] The cart and mules powerfully symbolized King's roots, his close ties to southern blacks, and the poor people's campaign, King's "own last and greatest dream."[66] Today the link between blacks and mules has much less symbolic power, and if the shackle between blacks and mules has been broken, then so has the connection between blacks and the land.[67]

White southerners bound mules and blacks so closely, and they had done so for so long, that it may have been necessary for blacks to leave the land in large numbers in order to break the mental associations held by whites. As long as blacks remained on the land, many white southerners saw mules and blacks tied together with all the connotations that their linked images brought to mind. In this area, the power of the mule's imagery may have been significant since southern blacks, cotton, traditional social relations, and mules were so closely intertwined. If the southern mind has a propensity for embracing myths, this was a particularly powerful and long-standing image that went to the heart of southern existence.[68] The myth and reality of the connection between mule and black was so strong that southerners, black and white, may have been caught up in it to the extent that they found it difficult to conceive of a South without mules and blacks at work in the fields. As the mule era ended, the number of African American farmers declined rapidly for many reasons, but among the most ironic was the push to drive blacks from the Mississippi Delta. Retaining black labor for cotton cultivation had long been a central concern of planters in the region, but when that labor was no longer needed, white planters wanted the region to themselves.[69]

As the twentieth century matured, mules came to symbolize more the negative aspects of tenancy and sharecropping. The view of the South as agriculturally and economically backward was not new as the twentieth century dawned, but the mule began to develop as a symbol—along with staple agriculture, gullied fields, and dilapidated tenant houses—of the region's cultural and economic poverty. In "Barn Burning," William Faulkner describes the benighted Snopeses' trek from one farm to another. Their wagon was pulled by a pair of "gaunt mules."[70] Faulkner could use poorly cared-for mules to reinforce the deprived state of the Snopes family because the mule carried with it many negative connotations, and because the condition of a mule spoke volumes regarding the economic condition of a family.

Reflecting the reality of Faulkner's interpretation, a Tennessee landlord painted a similar picture in his correspondence to secretary of agriculture

Henry Wallace. W. O. Armstrong described his tenants' condition in 1933: "The tenant's [*sic*] clothing was scarce, not any too much food, household goods badly worn, the little furniture shacked, no bottoms in their chairs, no cows, no pork hogs and a precious few chickens. *The mules of these poor laborers were poor, sore backs, sore shoulders, harness repaired with hay wire and the tools were scarce and badly worn.*" Praising federal programs that helped his tenants, Armstrong reported that tenant conditions began improving in 1934. By 1935, he noted of his tenants, "They have fat mules this year. No sore backs, better harness and better tools."[71] Armstrong's tenants may have been unusual in light of their healthy mules, for the *Southern Agriculturist* reported in 1935 that six out of every seven mules "seen in Georgia and Alabama were too thin to do their best work and furnish the longest years of usefulness."[72]

Mules often looked thin, but another reason for the negative image of the mule was the fact that it had a mind of its own. It could both help the farmer and hinder him in his work, unlike the machine which did the bidding of the farmer without balking out of pain, habit, a contrary nature, or simple perversity. With a mule, working in a field was at best a cooperative effort between man and animal. Whether laying off a field or plowing, according to Harry Crews, "it was much easier if the abiding genius of a good mule was brought to bear on the job." Some mules were so adept at field work "that a blind man could have laid off straight rows behind [them]." These types of mules were a pleasure to work, since "they brought a lovely exactitude" to whatever they were asked to do. But even the best mules could not completely offset poor plowing. Some fields had straight rows, and others seemed to lack any rhyme or reason. It was said that some fields were so poorly laid off that the farmer must have been drunk when he plowed it, and that he needed to get drunk again each time he cultivated the crop.[73]

A mule could complicate a farmer's life by harming him physically. Tractors and other machinery could likewise hurt a farmer, but the blame for being injured by a machine usually fell upon the operator, while a mule was often credited with premeditation. Even so, it appears that farming with draft animals was safer than farming with machinery. Even before full-scale mechanization, machines caused a higher percentage of farm accidents than livestock, and machines with engines were the most dangerous of all. In 1938, one study concluded that machinery caused thirty-eight percent of all farm accidents. Livestock caused 26.5 percent, falls 15 percent, and woodcutting 7 percent. The remaining 13.5 percent of accidents fell into categories such as falling objects, gunshots, explosions, fire, and poisons. Significantly, sixty percent of the machine-related injuries were linked to engine-propelled equipment such as automobiles, tractors, trucks, and combines. Importantly, eight out of every ten livestock-related accidents involved horses or mules. Farming was, and remains, an extremely dangerous occupation, but it may have been less dangerous before the advent of tractors and related machines.[74]

At the very least, as was recognized at the time, tractors were mechanical and did not have the ability to stop when the operator yelled at it. Tractors also had spinning power takeoff shafts that could catch loose clothing, engine cranks that could break arms, and flammable fuels that could catch fire or explode from a stray spark.[75] Speed, one of the most important benefits cited by tractor advocates, had a dangerous side as well. The National Safety Council reported that speed played a central role in most tractor upsets. By the late 1950s, approximately one thousand farmers were killed in tractor accidents each year.[76]

Draft animals possessed their own dangers for farmers. Mules were legendary for their willingness to avenge a perceived wrong. Kicking was perhaps the most damaging method, but they could employ more subtle means as well. Since mules were so careful about where they placed their feet, if a mule stepped on a handler's foot, one could be reasonably certain that the animal meant to do it.[77] The mule's reputation for surefootedness, along with his notoriety for revenge, would always make the farmer wonder whether the mule was purposeful or not in his offenses. Farmers acknowledged that mules often had a sound reason, such as being startled or abused, to lash out with their potent hind hooves, but there would always be some question if one could not directly connect a cause and effect. Because mules were alleged to be more combative, they protected themselves more readily than horses. The poem "A Pious Wish" alludes to this, as well as to the ill treatment many southern draft animals endured:

I knew a man who kicked a mule,
Offside, and called the beast a fool.
The mule kicked back; they swept that man
Up with a broom and dusting pan.
I know another man who bats

His bony horses in their slats,
And jerks their heads and calls them fools—
I wish his horses all were mules.[78]

Sometimes mules might have their own inscrutable reasons for kicking, as "Managing a Mule" attests. Published in an 1882 agricultural journal, the poem reflects the treatment blacks often endured. It seems from the dialect that the main character in the poem is meant to be African American, although it is not clear whether he is a landowner, tenant, or sharecropper. The thrust of the story, however, focuses on a common theme of mule handling. Just as the plowman thinks he has the animal under control, he is caught off guard:

> You Nebuchadnezzah, whoa, sah;
> Whar is you tryin' to go, sah?
> I'd hab you for to know, sah,
> I's a holdin' of de lines!
> You better stop dat prancin'
>
> You's powerful fond of dancin'
> But I'll bet my yeah's advancin'
> Dat I'll cure you ob your shines.
>
> Look heah mule! Better mind out—
> Fus' t'ing you'll know you'll fin' out
> How quick I'll wear dis line out
> On your ugly, stubbo'n back.
> You needn't try to steal up
> An' lif' dat precious heal up,
> You's got to plow dis fiel up,
> You har, for a fac.'
>
> Dar: dat's de way to do it!
> He's coming right down to it,
> Jes' watch him plowin' to it!
> Dis nigger ain't no fool!
> Some folks dey would beat him;
> Now, dat would only heat him;
> You mus' reason wid a mule.
>
> He minds me like a nigger,
> If he was only bigger
> He'd fotch a mighty figger,
> He would, I tell you! Yes, sah!
> See how he keeps a-clicken
> An neber tinks o' kicken!—

Whoa, dar, Nebuchadnezzah!
....
Is dis heah me, or not me?
Or is de debil got me?
Was dat a cannon shot me?
Hab I laid here more'n a week?
Dat mule do kick amazin'!
De beast was sp'illed in raisin'—
By now I s'pect he's grazin'
On de oder side de creek.[79]

Machines had no perverse inclination for revenge. Neither did they learn good or bad habits from experience, but they did have the potential to remake the rural South. Perhaps most importantly, machines such as the tractor were not part of the traditional southern agricultural system. The machine held the promise of a truly new type of South in the estimation of some experts. Arthur Raper noted in 1946, "The gains in mechanization thus promised may go far toward solving some of the South's old economic and social problems." Among the problems Raper linked to the "very lack of mechanization of cotton production" were "small farms, low incomes, child labor, irregular school attendance, poor housing, credit farming, excessive soil depletion, [and] inadequate medical care for a great proportion of the lower-income families."[80]

Employing the same logic as Raper, the USDA stressed per-capita production in its publications and explained that increased production by fewer persons meant that rural America was better off than ever before. Between 1910 and 1930, the number of agricultural workers decreased, but even so, "there was an increase in both acreage of harvested crops and total crop production. The increase in production per worker was due, in large measure, *to the increased use of power and machinery on farms.*"[81] Most of the increases occurred in regions other than the South, but the same formula of increased power and machinery, the USDA assumed, could be applied to the South with similar results. In this view, fewer people on the land, using more and better equipment, would result in a wealthier South. Gilbert Fite argues that "[t]o really solve the problem of rural poverty required reducing the number of people in the South who depended on agriculture for a living."[82] Without doubt, agricultural engineers and the tractor industry were most vocal and adamant about the benefits of mechanization. The positive results on agriculture of shifting from hand and animal power to machine methods were so obvious to engineers that to reflect on them would be akin to "carrying coals to Newcastle."[83]

Not everyone agreed with the USDA's prescription for southern improvement, however. The Agrarian critics who produced the classic *I'll Take My Stand* saw much to be commended and defended in the rural South. But, to the Agrarian philosophy, the mule was not simply a quaint element of the traditional South. The Agrarians' attack on industrial agriculture contained chords from the draft animal–tractor debate, but it took place on a different, symbolic level seldom touched upon by the more practical-minded draft-animal proponents of the Horse Association of America. To the Agrarians, the draft animal stood as the last barricade against the machine age, against the tractor, and against the death of a vital rural culture which, if not totally self-sufficient, could at least feed its own work animals, even if it could not always produce them.

Even as the mule began its decline as the primary source of farm power on southern farms, a philosophical debate with the mule and machine as the key symbols developed over the consequences of adopting machinery on southern farms. Viewing the massive mechanization in the Midwest during the 1920s, the Vanderbilt scholars asserted that a similar shift in the South held profound implications for the region's mule-powered agriculture. The question, then, to critics and opponents of tractorization and industrialization of southern farms hinged not upon the speed of work, but upon other, somewhat less tangible—but, in their minds, even more significant—factors. Andrew Nelson Lytle treated this point in his contribution to the Agrarian manifesto *I'll Take My Stand.* Lytle believed convincing the farmer "that it is time, not space, which has value," would be difficult, if not impossible, as long as the farmer understood that he could not control time, but that he could "wrestle with space, or at least with that particular part which is his orbit." One of Lytle's farmers summed up the whole debate when he stated, "'as soon as a farmer begins to keep books, he'll go broke shore as hell.'"[84]

At the same time, industrial America offered a rapidly expanding number of labor-saving devices and a host of modern accoutrements that promised to make farm life easier as well as more enjoyable. Modern technology promised rewards nearly beyond description. The National Association of Farm Equipment Manufacturers asserted in 1925 that "every farmer should be able to afford the good things that life offers," such as an automobile, a radio, a healthy savings account, electric lighting, a telephone, and a water system. The farm wife should be included as well. Her kitchen "should have every convenience" to lighten her workload. The farmer's children were entitled to "the best education there is." Finally, the farmer "should have time and wealth left after his work is done to enjoy his earned leisure."[85]

Industrial spokesmen were not alone in their representation of the tractor as salvation for farmers. Herman Clarence Nixon, for example, had contributed to *I'll Take My Stand* but was not satisfied with the direction of the Agrarian argument after the book's publication. He sought a broader approach to solving the South's rural ills than what he believed the Agrarians advocated. He was not blind to the potential problems society faced by relying on machines, but he believed that machines, properly harnessed, could solve the South's problems. Nixon went so far as to assert that "leisure-loving southerners" would be especially appreciative of "mechanical slaves" since the region had experienced human slavery. Unlike antebellum slavery, however, the emerging era of mechanical slavery promised to spread its benefits among many more people. "With machines for the slaves of a democracy," Nixon reasoned, "we may be enabled to release our minds from concerns over economic production or income and to indulge in the non-economic activities and contemplation of life, just as did a few privileged Southerners in the eighteen-fifties." Nixon viewed the future with a hard-edged realism. "The mule in the flesh must be adjusted to the steel mule, and both must be subordinated to the service of rural society," he wrote. Even with his pragmatic approach, Nixon hoped that the southern machine age would produce something of a utopia in the region, just as the Agrarians believed that a nonmechanized South would provide the best possible society.[86]

The issue of mechanization, then, was as much philosophical as it was mechanical. The lines of debate stood clear before the cotton picker signaled the end of the mule era. At a 1936 Mississippi cotton-picker demonstration, two "prominent men in the cotton growing business" assessed the potential impact of the new machine. One opined that the cotton picker would lead to "more profit and greater opportunities for the average Southern farmer and his family." The other observer noted that a viable cotton picker would "be the death knell for family sized farms and for tenants. It will encourage and develop large holdings."[87]

Some factors tended to limit mechanization and other forces pushed planters to depend more on modern equipment and methods of growing cotton, but in the end, mechanization deeply affected landlord and tenant relations in the cotton South. Changes in the labor supply ranked high on the list of reasons plantation operators in the Yazoo Mississippi Delta gave for mechanizing. The labor issue cut across the subregions of the South and across the decades. The late 1930s echoed with a strong outcry by landowners over the unwillingness of tenants and sharecroppers to perform agricultural labor. One South Carolina farm wife complained directly to President Franklin Roosevelt. She explained that because of government jobs and programs, "the absence of farm labor in this section of the South has become quite serious to the farmers." Although farmers offered laborers higher wages, small plots of land for their own crops and gardens, and other incentives, she complained that laborers refused to accept the enticements. Some farmers had "threatened to sell their mules, as they cannot make a crop without help, which cannot be had in this particular section." Mrs. Walling concluded that the farmers were "glad to have had the negroes work on Government projects during the winter months to make extra money; however, it is now high time farm work is started in this section and the farm work will be permanent from now on through harvest time."[88]

As far as most landowners were concerned, the labor situation continued to deteriorate. In the years following World War II, planters who had mechanized and those who still used mules complained about a general lack of workers. Mechanized operators deplored the dearth of skilled workers, especially tractor drivers. Nonmechanized planters complained that there were too few seasonal workers and sharecroppers. Many planters pointed out that a major part of the problem was workers' refusal to accept the going wage rates, and planters often blamed federal programs, ranging from the WPA during the 1930s to soldiers' allotments, aid to dependent children, and old-age payments.[89] One Delta planter reported an "unfortunate" state of affairs concerning the "loss of womanpower" on the plantation he managed. Monthly payments to soldiers' wives and children made it possible for some potential farm laborers to move to town and do odd jobs, light work, or domestic labor.[90]

Most planters believed that labor was growing less willing and less dependable, although they also acknowledged that workers seemed more capable in the early 1950s due to more education and increased experience with machinery. Plantation operators consistently blamed influences outside their own plantation for labor problems—neighbors offering cash "furnishing," better housing, electricity, access to roads, and other perquisites. Sociology professor Harald Pedersen, who examined post–World War II mechanization in the Yazoo Mississippi Delta, believed that the attitudes planters held toward the changing labor situation tended to "push him into accepting techniques that he does not understand and in which he has little faith, but which are effectively moving him toward more complete mechanization of operations."[91]

Pedersen studied the dynamics of mechanization closely, but the more common response was typified by the president of the National Cotton Council, who observed that while there were both mechanical and social problems related to mechanizing the cotton routine, "we'll gradually work them out as we go along. It will take time. I don't look for any sudden or serious displacement of people. There hasn't been anything serious yet, and there's none in sight."[92]

Tractors led to significant tenant displacement, but just as significantly, they changed the relationship of landlords and laborers, heralding a revolution in the way the rural southern society would be organized. Even so, it is extremely difficult to generalize concerning rental and cropping arrangements in the South, especially once tractors became part of the arrangement. One study of southern piedmont farms found no fewer than 65 different arrangements between landowners and 185 renters. The study further noted that tractorization had been slow in the piedmont, mainly because of sharecropping. As the report observed, "One of the basic assumptions underlying the overall study is that mechanization in the Southern Piedmont has been slowed down or hindered by the traditional rental arrangements of the area." Cotton and mules fit well within the traditional labor system of the South, and to change one aspect of that tradition was to set into motion a myriad of other, perhaps unforeseen changes.[93]

Although generalizations concerning the impact of tractors on tenants are difficult to make, it is clear that new relationships had to be forged in the face of new technologies. In piedmont Alabama, Georgia, South Carolina, and North Carolina, "the introduction of tractors into cotton farming . . . tended to upset the traditional systems of dividing receipts and expenses of the cotton crop." Not surprisingly, comparisons of rental arrangements showed that renters using all mule power farmed under the same basic arrangements whether they worked on farms with or without tractors. At the other extreme, wide variations could be found in arrangements on farms where renters used tractors not belonging to them or where tractor "custom" work was performed for the renter. In the latter situations, differing expectations of landowners and renters could create new levels of tension and lay the foundation for future misunderstandings.[94]

Tractors altered much more than just the way the ground was plowed. Mules represented and supported not only a traditional means of farming, but a long-standing type of relationship between landowner and tenant and, in many ways, long-standing modes of race relations. Few tenants could purchase tractors. Thus, if the landowner purchased a tractor and insisted the tenant use it, the tenant lost on two counts. First, he lost the bargaining power that providing his own work animal afforded him. Second, his mules became obsolete overnight and the animals' value dropped accordingly. Many tenants found themselves as day laborers, working for the landowner in a profoundly different way than they had worked for him previously. Generally, tractors put tenants and sharecroppers at a decided disadvantage, and as time passed and more farms mechanized, options narrowed for those who did not own farmland. The replacement of mules by tractors also removed a living creature from the complex landlord-tenant relationship and replaced it with a machine. As government programs provided some limited aid to tenants and sharecroppers during the 1930s, and as machinery dehumanized the southern plantation, the landlord's sense of responsibility toward "his" tenants and sharecroppers steadily declined.[95]

Tractors also affected the way landowners and tenants shared expenses, although the arrangements varied from farm to farm and individual to individual. Some farmers, for example, used tractors only to prepare the seedbed. In such a situation, the landowner sometimes did the tractor work and allowed the cropper to work a set number of days. In other cases, landowners charged the cropper per acre for tractor work, often slightly less than the local rate for custom tractor work. The latter alternative created serious problems for sharecroppers since they had only their labor to market. By charging a cash-per-acre price for tractor work, the landowner at once reduced the amount of work available for each cropper to perform and increased the cropper's cash indebtedness. The landowner might have wage labor for the cropper or, in a few cases, the cropper's production might increase enough to offset the larger cash demand the new arrangements imposed on him.[96]

Landowning planters had definite views toward tractorization, and those views, in concert with a host of other factors, influenced the general southern move toward a mechanized agriculture. On the other hand, tenants also had opinions of mechanization, although those views were less influential and less well known. Tractorization affected tenant and sharecropper life profoundly. When planters bought and used tractors on their lands, the work required of tenants declined. Once, tenants had begun plowing in the spring and stayed busy with various duties until the cotton crop had been picked late in the year. With tractors doing the heavy plowing and sometimes the planting, tenants might not be needed for chopping until the crop was well along or "green in the row."[97]

Most tenants and sharecroppers could grow more cotton per acre in 1950 than in 1930 due to improved varieties of cotton, better farming practices, and better fertilizers, but this did not necessarily offset higher costs. In the southern piedmont, 1947–51 yields averaged one-third higher than 1930–34 yields. The average 1928–32 cotton yield equaled approximately 174 pounds per acre. During the World War II years, the national average hovered around 260 pounds of cotton per acre. In terms of feed crops, however, grain production which demanded little hand labor was mechanized before cotton culture. Landowners, therefore, raised grain themselves and sold it to their tenants. In effect, the tenants paid the landowners to grow grain for them on a custom basis.[98]

In some areas where off-farm employers competed with landowners for laborers, and where factories and farms were fairly close together, tenants might work in both agriculture and industry. In the southern piedmont, landowners had to share laborers with factories, but found that workers often neglected the cotton crop, despite having the use of a tenant house, garden, pasture, and firewood at little or no cost. Landowners responded with a "hoe-cropping" arrangement made possible by the tractor. The renting family provided labor to chop, hoe, and pick the cotton crop, while the landlord paid all expenses and often prepared the land, planted the crop, and cultivated with tractor power. For hoeing and picking, the renter received one-third of the crop. It appeared to be a workable solution economically, but advocates perceived other benefits for renting families since many "families are better adjusted to life in the country and prefer to remain on the farm than move to town and face the somewhat difficult period of adjustment to a new way of life. Through this arrangement he becomes in effect a 'city worker' with a 'home in the country'—a status enjoyed by comparatively few of his counterparts in other sections of the United States."[99]

Relatively few tenants had the option of working in a factory and living on a farm. Most remained in a truly rural environment and had to choose from among limited options. In the Yazoo Mississippi Delta, one study found that most tenants preferred working on mechanized plantations for two main reasons. First, tenants believed work could be done more quickly with tractors, which translated to fewer hours in the field. Second, the duties required of croppers on tractor farms were easier. Tenants explained, "A man can sure do a lot of walking behind a mule." A significant number of tenants preferred mule plantations, however, although their reasons were not as consistent. Importantly, tenants who preferred mule plantations did not necessarily prefer mules over tractors; they simply did not care for certain implications of working on a tractor farm. Indeed, in Harald Pedersen's survey of the Yazoo Mississippi Delta, not one tenant "mentioned the presence or absence of tractors as a determinant in his choice of a plantation." Factors such as the reputation of the manager or owner, the type and amount of "furnish" provided, the condition of the house, the kind of land, and the treatment of tenants ranked high in importance for tenants choosing a plantation—not whether mules or tractors were used.[100]

Still, tenants and croppers had views on tractorization. Perhaps most significantly, many tenants wanted no part in learning to drive, service, or repair a tractor. Some tenants chose to migrate rather than face mechanization. As World War II intensified and the postwar economy boomed, the military and industrial demand for manpower pulled many agricultural laborers off the land and out of the South. Increasingly, tenants defined opportunity not in terms of learning more skilled farm jobs, but in terms of migration off the land. In turn, this pressured landowners to seriously consider farming with tractors, even if they did not completely understand or trust the new machinery and its demands for new farming methods.

In general, Pedersen found in his study of the Mississippi Delta that most laborers on plantations were not keen on tractors. Perhaps most significantly, tractors changed the relationship between management and labor. Tenants did not welcome the increased supervision that usually accompanied tractor use during the early stages of tractorization. On a large plantation that worked four or five tractors, a supervisor or foreman would be standing at the end of each row watching the work progress and "looking at them—all day." The newness of the technology and the large investment planters had in tractors pushed them to supervise closely what the machinery was doing. Traditional farming methods meant longer hours and more physical labor, but Pedersen found among many laborers he interviewed a preference for the less strictly observed system of cotton production. If croppers or tenants remained on the land, Pedersen concluded that "they prefer the traditional role of cropper or day laborer to the more closely supervised semi-skilled or skilled occupation of drivers."[101] Thus, while Delta plantation sharecroppers were, in James Cobb's words, "its most tightly supervised farm workers," tractorization meant that supervisors maintained an even more vigilant watch over their charges as the plantation edged closer to an efficiency befitting industry.[102] Ironically, the adoption of new technology led to a return to supervisory techniques akin to those used under the slavery regime.

Workers, then, confronted a difficult decision as long as some planters continued to farm with mules. They could stay on a mule farm and be physically taxed to the limit, or they could find work on a tractor farm and endure close supervision and a new relationship with their employers. Once tractors had replaced mules entirely, however, workers could not leave a tractor plantation for a mule plantation. Tractors also made the climb from cropper to tenant nearly impossible for southern sharecroppers. Before mechanization, the purchase of a mule stood between sharecropper and tenant status. By the late 1940s, a mechanized, family-size farm in the Mississippi Delta which grew 150 acres of cotton operated nearly $11,000 of machinery and equipment dedicated to cotton alone.

The total investment in machinery rose to more than $16,000 with the addition of 175 acres of corn, oats, or beans to the equation.[103] In Alabama, Georgia, Louisiana, Mississippi, North Carolina, and South Carolina, more than one-third of all black farmers owned no tractors, horses, or mules at midcentury. One-quarter of white farmers fell into the same category. Tractors increased the cost of the transition from sharecropper to tenant farmer many times and thus tended to solidify class lines while decreasing the competitive ability of poor southern farmers, black and white. Economist Victor Perlo argued in the 1950s that "[i]n the South of the United States, in the middle of the twentieth century, hundreds of thousands of poor farmers, mainly Negroes, are no better off for equipment than the horseless peasants of tsarist Russia, the lowest landed strata in the backward agriculture of that epoch."[104]

Policy makers had long recognized the difficulties of converting sharecroppers into independent farmers, particularly at low cost. The success of resettlement and rehabilitation programs may have depended upon the continued use of low-cost farm power in the form of mules. Established during the 1930s, these programs provided for the purchase of acreages too small to support a tractor and funded on a scale suitable only for the relatively low-cost mule. The Brookings Institute's Frank Tannenbaum, who studied southern agricultural and social problems, advocated a comprehensive land settlement program in the 1930s with the objective of converting tenants and sharecroppers, "and thos[e] recently set afloat, into an independent small landowning agriculturist as quickly as possible, and upon a *cost basis that will be low enough* to make it feasible for this submarginal group in our population to secure, gradually and over a period of time, the independence derived from unencumbered land owning."[105]

The high cost of tractors would have required prohibitively high funding levels from the federal government. Government programs such as the Farm Credit Administration helped many tenants and sharecroppers, but the impact of the Depression on rural southerners stymied aid efforts. According to one estimate, with farm tenancy increasing between 1880 and 1935 at the rate of 33,465 farms each year, a $4,000 investment per farm would have cost $133,860,000 per year just to take care of the increase in tenancy.[106] Some families such as the Langleys of Alabama received nearly $4,000 in the form of a loan.[107] Other farmers, under the auspices of the Resettlement Administration, garnered smaller amounts of money. Mississippian John Jones was placed on a forty-five-acre cotton-cane-corn farm and received $222 for food, clothing, seed, and fertilizer, and $113 for a mule, a cow, and farm equipment. Jones made some headway, with close supervision, toward self-sufficiency, but neither he nor government officials expected him to move beyond that point.[108] Programs set up for small farmers in a largely nonmechanized decade, therefore, may have been doomed to fail as tractors and other machinery transformed the South. Community ownership of machinery was not unknown, but farmers most often wanted their own land and equipment.[109]

Mechanization proponents promised that farm life could be as fulfilling and profitable as urban life and, in doing so, demonstrated an urban bias in the underlying assumptions about the relative fulfillment of industrial and rural life. Nor was this view new to the twentieth century. Approximately halfway through the nineteenth century, machine advocates pushed "to regularize farming and get agriculture up to snuff."[110] Around the same time, an important shift occurred in agricultural machinery development as well. Prior to 1840, machinery improvements in the agricultural realm led principally to improved culture, whereas after midcentury, improvements tended to focus upon labor-saving devices and increased production through more extensive farming methods.[111] Greater speed and less labor, in other words, became a central measure of quality farming as the twentieth century approached. Mules and mule-powered farming did not fit well into this paradigm.

Conclusion

In many ways, to "sing the saga of the mule," as William Faulkner urged, is to relate the story of the South itself. For decade after decade, the mule stood as one of the most visible constants in southern rural life, as well as one of the most conspicuous reminders that the South was "different." "There must be mules or cotton cannot be raised," observed one author in the early twentieth century.[1] But that did not hold true over the long haul, and mules elicited much comment when they began to disappear from cotton fields. As a writer in the *Progressive Farmer* observed in 1958, "This seems certain: The mule may be fading from the Southern farm scene, but he is not losing his place in the hearts of readers."[2] Like many commentaries on the demise of the mule, this one suggests that absence may have played a large part in producing fond memories of the animals and the way of life they connoted.

The story of the mule's passage through southern history reflects the region's passage from a slave society into tenancy and sharecropping, and then into an era characterized by rapidly expanding urban and suburban areas, capital-intensive agribusinesses, and a depopulated countryside. In the broadest sense, between 1850 and 1950 the mule South evolved into the tractor South. When the census no longer deemed mules worth counting in the early 1960s, the end of an era truly was at hand. The foundations of the old order, weakened by myriad changes, would sustain the long-standing economic, political, and social structure but a short time.

When mules began to disappear after World War II, there were perhaps a few tears, but most eulogies were lost amid the cacophony of farm trucks, tractors, cotton pickers, and "a long-awaited economic 'take-off' replete with dynamic industries and a rapidly expanding middle class."[3] By 1950 the South was beginning to be characterized as much by the urban skyline as it once was by the rural fence line and furrowed field. Mules and the agricultural system they powered disappeared in the transition, and as in all such changes, there were gains and losses as a result.[4] Whether the gains overshadowed the losses is a matter of perspective, but it is certainly a question worth pondering, especially in terms of how society chooses to use technology and to what uses it is put.[5] A few observers have analyzed the changes wrought by technology and attempted to reassess what exactly was lost and what was gained in the shift to a capital-intensive, mechanized agriculture and a new economic base, as well as what implications the changes hold for the future of the region now that we have entered the twenty-first century.[6]

Mules, of course, were first and foremost work animals. But they also reflected something of the character of the region, as well as the worldview of those who bought and used them. The mule represented innovation of a sort for antebellum planters, but bondage for those who used the animals. After the Civil War, for freedmen the mule was an essential element in the elusive quest for economic independence, a hope echoed in the desire for "forty acres and a mule." In time, however, the mule symbolized sharecropping's Sisyphean judgment upon southern farmers, black and white, and upon the region as a whole. By the 1950s, to critics of the region, the mule embodied everything negative about southern rural culture and agriculture, and in the post–World War II boom, many southern farmers optimistically abandoned the mule for the tractor and the cotton picker.

Many southerners left farm life voluntarily for the lure of jobs and wages in cities, but many others were pushed off the land because machines and new farming techniques replaced hand labor. Landowners, especially planters, found that the labor they had desperately needed for so long was leaving, but they also found in that exodus that labor, both human and animal, could be replaced by machines.

The South, as C. Vann Woodward consistently intoned, is a land of paradox and irony.[7] In that vein, it is fitting that the mule can represent both the worst and best aspects of the region. Even as it became linked in the popular mind to monoculture and poverty, the mule also represented the strength of a rural, agrarian culture which survived civil war, reconstruction, and depression.[8] Both white and black farmers used millions of mules, but black farmers were closely linked to mules in the southern white mind, and mules carried definite class and status connotations. Mule use was also ironic in that the various subregions of the South that employed mules to the largest extent did not breed the animals, but purchased them from outside suppliers. An intriguing and ironic fact

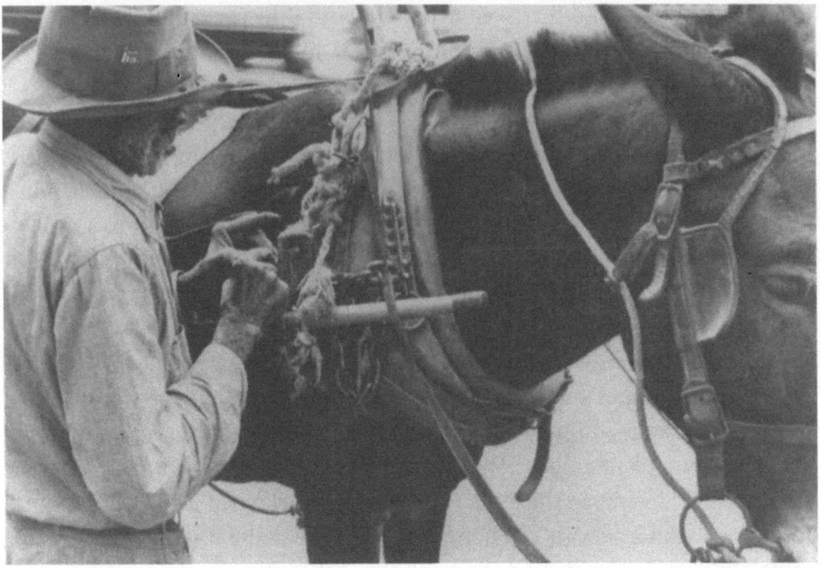

Fig. 12. African Americans and mules were closely linked in many ways and on many levels throughout the era of mule use. (Courtesy of Mississippi Department of Archives and History)

is that the most traditional of regions would employ an "unnatural" animal that would come to represent some of the most traditional elements of southern culture.

Without a doubt, the South looks more and more like the rest of the nation, and vice versa, with each passing day.[9] The siren song of economic success and material wealth had long called to the South, but in the latter half of the twentieth century, the region was finally able to steer in the direction of the music instead of being held away from it by powerful currents. Instead of being dashed to death on the rocks, however, the South's fate apparently is to become like the rest of the nation.

The changes produced by the swirls and eddies of modernization affected every southern state to a greater or lesser extent, but Georgia is typical of what happened in the transition from mule to tractor South. In 1930 the state had a population of 2.9 million persons, of whom 1.4 million lived on farms. In 1990, out of a state population of 6.4 million, just over 80,000 fell into the farm population classification. The state's farm population decreased from 48 percent to 1.2 percent in sixty years. Even more dramatically, these changes deeply affected blacks. In 1945 black farmers comprised nearly one-fifth of farm own-

Fig. 13. Although mules are often associated with African Americans, mule use cut across racial boundaries as this Lee County, Mississippi, farmer shows. (Courtesy of Mississippi Department of Archives and History)

ers in Georgia. But in 1990 the number of black farm owners hovered around three percent of all farmers, with more women farm owners living in the state than black farm owners. Black farm operators increased slightly by 1997, but still comprised only 1,329 of 40,334 farmers in the state.[10]

The fact that black farmers disappeared along with mules is more than coincidence. The forces that led to the disappearance of mules from southern farms pulled and drove blacks from field to city. Nationwide, the 1990 census found fewer than twenty thousand black farm owners, a decrease of some thirteen thousand from 1982. That number increased as of 2002 because new techniques were used to enumerate small farmers, but African American farmers still made up only a tiny percentage of the two million or so American farmers.[11] In the late 1980s, black farmers lost land at more than twice the rate of white farmers.[12] Southern cities absorbed rural people, as well as immigrants from other areas of the nation. Atlanta, Georgia, for example, boasted a 2003 metropolitan area population of more than 4.5 million people, and projections call for that number to continue to grow.[13]

If southern cities have changed enormously, so has the southern countryside. Among the most visible changes evident in the post–World War II South, then, are those connected with the decline of rural southern culture and agriculture, and the rise of southern urban life. Myriad contrasts exist between the mule South and the tractor South—noise, for example. The mule South was not necessarily quiet, but its farm noises for much of the early twentieth century largely came from living things. Cotton gins, steamboats, and railroad locomotives sent their sounds across parts of the South, with the railroad playing a preeminent role in Reconstruction and later in southern development, but cotton fields often only echoed these noises without adding their own sounds beyond those of living creatures, plowshares scraping the soil, and hooves plodding over clods of dirt.[14]

In one sense, then, the source of noises and the noise level itself were indicators of change and of new relationships that developed among men and machines over the course of the South's journey farther into the mainstream of American economic and social life. But it often meant fundamental change. Southerners for so long depended on human and animal labor to carry out the most tedious chores associated with staple agriculture that when the machine came, all of the old, often perversely comfortable, relationships that had accreted and ossified over the previous century had to change, too. Fundamental change occurred, but while the machine certainly removed many types of tedium, it replaced them with others.[15] Few farmers talk to or about their tractors the way they talked to and about their mules. Indeed, the mule's hearing is said to be so sensitive that, while one can talk to a horse, "you can chat and whisper to a mule."[16]

Account books displaced farmers' almanacs; scientific advice from the county agent replaced respect for traditional knowledge and natural signs; weekly and then daily trips to town replaced long days on the farm; and petroleum and engines replaced corn, hay, and muscle as the energy that powered southern farms. And as southerners moved off the farm, they lost contact with the culture surrounding the draft animals that had plowed the region's fields.

The most pointed and antagonistic critics of modern agriculture often focus on the United States Department of Agriculture, and to some that agency has come to symbolize everything that is wrong with agricultural and rural America today. That farm equipment manufacturers trumpeted the benefits of their wares without regard for the social repercussions of mechanization is perhaps unfortunate, yet understandable, according to this school of thought. But the federal government stands without excuse for accepting the same tenets as the manufacturers and encouraging farmers toward the same ends. Empiricism and quantification, in the form of ledger sheets and depreciation tables, almost by definition gut Jeffersonian ideals of an independent yeomanry. Pete Daniel asserts that "[w]ith the premise that science, mechanization, and large-scale farming provide the only model for U.S. agriculture, the USDA will continually attempt to restructure rural life into its updated fantasy of giant machines, robots, experts, computers, and chemicals that leaves no place for farmers who move in natural cycles."[17]

One must keep in mind that the view of agriculture as akin to any business or industrial undertaking predates the formation of the USDA. The USDA, argues another school of thought, merely supported endeavors for which farmers desired aid. As two officials explained in the 1930s, the unique conditions of American society "brought forth a constant demand on the part of farmers for labor-saving implements and machines and have furnished a powerful incentive to inventors and manufacturers for improving and developing machines."[18] The *Southern Agriculturist* noted in 1940 that "no thoughtful person would contend that the farmer should use ox-cart methods to compete with streamlined industry in this age of efficiency and keen competition."[19]

If this is true, the trend toward speed and size in agriculture may reveal much more about American society than it does about the Department of Agriculture, and it represents the triumph of industrial ideals over agrarian tenets in a broader context than the South or the American farm.[20] The Department of Agriculture, after all, took its official form in 1862 during the Civil War, a conflict sometimes interpreted to be a struggle pitting modernizing forces against traditional. Yet, extension agents also promoted home production of mules, which in some ways ran counter to their assumptions about the efficacy of labor-saving improvements, and pronouncements concerning the nature of the changes in rural society point to the existence of agrarian ideals even within the USDA.

J. A. Evans spent a long career in extension work. He had worked with Seaman Knapp, among others, and had been in farm demonstration work from its origins early in the twentieth century. Evans looked back over his years in agriculture and was pleased for the most part with the changes he had helped make. Still, he worried about the impact of modern life on rural culture and the nation. Sounding much like an essayist in *I'll Take My Stand,* Evans recalled in the 1940s that "urban culture in this age has conquered rural culture. We are money mad and thinking only in terms of money and what it will buy. People are living for today only." Even farmers overlooked the "enduring values and satisfaction of life" as they sought "to obtain at once and at any sacrifice all the alluring and desirable or useful 'gadgets,' the possession and use of which go to make up what we term our high standard of living." Evans called for a "rural culture, based on thrift, the love of home and the soil and the ambition to conserve, develop and beautify our farms as a precious heritage to be passed on to our children."[21]

What Evans did not acknowledge, however, and probably did not see, was the role that the USDA had played in creating the problems to which he referred. Through the USDA's emphasis on the need for farmers to increase efficiency, focus on business-type accounting methods, and achieve a standard of living comparable to that of urbanites, those virtues to which Evans alluded were eroded or displaced. This is the precise point on which critics of the machine focused in their defense of traditional southern culture. There was no way, in short, to possess the fruits of modern urban culture without altering the roots of the old, traditional rural culture. The paradox is, of course, that some elements of that traditional culture needed to disappear.

What is intriguing about the view held by Evans, apart from its agrarian themes, is that it and many of the programs advocated and carried out by the USDA actually paralleled the agrarian vision. Not only did agrarian symbolism fuel opponents of the USDA, it also provided the impetus for some programs within the USDA, particularly those in the South during the Depression, the decade on which the harshest critics of the USDA focus. Frank Owsley, one of the Agrarians, attempted to enumerate a detailed agenda for southern renewal in his 1935 article, "The Pillars of Agrarianism." He blamed "Industrialism" for driving the farmer "to accept industrial tastes and standards of living," and in the process lose his farm. The result was a countryside littered with "old cars, dangling radio aerials, rust-eaten tractors, and abandoned threshing machines and hay balers."[22]

Rehabilitating the population living on the soil was Owsley's first prescription for ameliorating the southern condition. Second, he argued that the soil needed a similar rehabilitation. Third, Owsley believed that "[s]ubsistence farming must be the first objective of every man who controls a farm or plantation." Yet he recognized that larger southern farms would be necessary. To Owsley, eighty acres seemed appropriate. Owsley did not see the USDA or scientific agriculture, at least in 1935, as opposed to his scheme. Farming, "if carried on with the scientific knowledge available today," Owsley noted, would benefit the South.[23] Importantly, he argued that scientific farming and draft animals meshed well. A healthy subsistence farmer needed—apart from sixty acres of cropland and housing—twenty acres of pasture, two mules, and two milk cows. Through live-at-home programs, home mule production projects, and similar endeavors, the USDA and its extension agents actually carried out parts of an agrarian agenda. At the least, then, one is faced with highly varied and complex relationships among the rural South, the Department of Agriculture, the land grant system, agricultural corporations, and agrarian and industrial ideals.

Looking back at the broad societal forces at work, one scholar has asserted that a number of factors influenced tractorization and wholesale mechanization. According to Wayne Rasmussen, during the post–World War II years, high prices, rising demands for more food, a shortage of farm labor, and government calls for increased agricultural production combined to lead farmers to embrace technological advances.[24] If the great number of American farmers preferred to drive a truck rather than a wagon, or to sit behind a tractor's steering wheel rather than walk behind a horse or mule, then the USDA was responding to the desires of a large number of its constituents, rather than dictating an unpopular public policy.

The languid pace at which much of the cotton South mechanized, relative to the rest of the nation, may reflect the fact that southerners continued to be partly out of step with the national consciousness until the Depression and World War II catalyzed profound regional change.[25] Certainly Jim Crow in all its forms deeply influenced choices people made regarding mechanization, and the drag on innovation because of race and labor questions must not be neglected.[26] In the broadest sense, the USDA did not have to convince farmers of the machine's benefits. "Today—you and your family can step into another world," asserted one 1940 advertisement. "From the gray, drab existence of mule farming, you can emerge into the sunshine of Better Living with Allis Chalmers power."[27] Many readily accepted the machine and its attendant baggage when it made sense to do so, and they easily moved into the machine age. Others succumbed to the tractor's promises of prosperity and relative ease, only to be rudely awakened by the realities of what they had done as payment after payment came due. "Keeping up payments on farm machinery," commented one observer, "has taken years off the lives of a good many farmers."[28]

By the mid-twentieth century, the mule no longer symbolized the future, or even a hardy rural culture. It became the opposite—a symbol of southern backwardness. During the Depression era, the economic plight of southern sharecroppers linked forever in the American mind the mule and a dysfunctional agricultural society. The tie between a healthy yeomanry and draft animals was sundered and replaced with the idea that a healthy agriculture rested upon machines, not upon mules. Thus, by the 1930s, the mule had begun to shift from a symbol of progress, which it had been during the antebellum years, to an icon of the South's perennial agricultural dilemma—symptomatic of the region's ills to modernist critics.

When southerners accepted a definition of mules as backward, they rejected the animals quickly. Sweat, toil, and physical labor, of man or animal, was perceived increasingly to be backward, especially when a machine was available to supplement or replace man or beast. In the end, the mule's passing signaled a crucial and perhaps final surrender to a different set of standards, but there was always some question regarding the future, at least among some. Donald Davidson captured the southern ambivalence toward the modern world perfectly. Writing of the postbellum era, Davidson suggested that although the southerner was agrarian at heart, "he had been forced to wonder whether the ingenious Yankee might not be right after all." "Thus he remembered the faith and hankered after the fleshpots at the same time. But industrialism, declining to be treated as a mere hedge, began Sherman's march to the sea all over again. It piled ugliness upon wreckage and threw the old arrangements out of kilter. The United Daughters of the Confederacy and the Kiwanis Club flourished side by side. Mule wagons and automobiles, fundamentalism and liberalism, education and illiteracy, aristocratic pride and backwoods independence disproved the axiom that two bodies cannot occupy the same space."[29] Likewise, mules and tractors coexisted, but only for a time. As mules yielded to tractors, agribusiness supplanted agriculture.

Mule farming differed in important ways from tractor farming, and heavily capitalized agribusiness differs significantly from traditional farming. The tractor farmer invested heavily in his equipment and depended on outside fuel sources, while the mule farmer invested comparatively little money in his stock and equipment, and raised at least part of his animal's feed on the farm. On the other hand, southerners were used to buying farm power outside of the farm since most mules were not produced locally. The differences in costs of mules versus tractors were substantial, however. Even today, American farmers who use draft animals insist that their type of agriculture succeeds because of the relatively low levels of investment required. But quantitative differences reflect only part of the story. Agrarian critics focused on the rest of the equation. Mule farming and tractor farming were and are qualitatively different.

To mechanization's opponents, modern, scientific agriculture represented the destruction of everything they held dear, the annihilation of the very components that set the South apart from the industrial world. While Agrarianism may have provided some of the most eloquent and pointed criticisms of southern progress, other less erudite commentators expressed similar misgivings about tractors. In the 1930s, one tenant summed up his feelings succinctly: "The Farmall is ruining our country."[30] This attitude summarizes what scholar James Montmarquet describes as an inherently agrarian response to mechanization. From the agrarian perspective, qualities such as "honesty, simplicity, independence and plain hard work" are victims of technological progress. Given this, Montmarquet asserts that "agrarianism will increasingly view technological progress and its consequences with alarm."[31]

According to the view developed by the Vanderbilt Agrarians, the triumph of the tractor on southern farms meant that the contrast between the South and the rest of the nation would dim, and the positive elements of southern culture, so bound up with its unique agricultural system, would be lost. But supporters of the machine held no such view. In Arthur Raper's view, the South was out of tune with the nation at large. The South produced its staples largely by hand, but the same crops were "processed upon leaving the farmer by big machines and mass production methods, *which are in harmony with the rest of the national economy.*"[32] This point quickly raised the ire of modernization's critics.

Donald Davidson perceived such comments not as substantive criticisms but as groundless taunts that were "intended to discredit the traditionalist by stigmatizing him as a traitor to an idea of progress that is assumed as entirely valid and as generally accepted." Critics of the traditional South, in Davidson's view, hoped "to poison the traditionalist's own mind and disturb his self-confidence by the insinuation that he is a laggard in the world's great procession" by making his "faith in an established good" seem like nothing more than "nostalgic devotion to a mere phantom of the buried past. His opposition to the new—no matter how ill-advised, inartistic, destructive, or immoral that new may be—is defined as a quixotic defiance of the Inevitable."[33]

Davidson's "Inevitable" did happen. To the generation of southerners who grew up in the 1940s and 1950s and witnessed the decline and death of the mule South, the disappearance of the mule itself held significant meaning. Doubtless influenced by nostalgic memories and myopic recollections, individuals such as Joshua Lee who wrote about the mule South correctly focused on the decline of the mule as symbolic of the end of traditional southern culture. Writing decades after his return from World War II and the death of the farm's last mule, Lee asked what this death meant, and then eloquently answered his own question:

> Why was I so upset over Old Alec's passing? True, he had been very special to me, but so had others of our horses and mules, and I'd not had nearly so many tears for them. Perhaps it was because Alec, more than any

> of the others, symbolized for me the old cotton South that had spawned me and nurtured me to manhood. At that very time the vast technological and social changes generated by the war augured the demise of all Alecs and the way of life they supported. Thus I suspect that my tears were being shed not so much for that contrary old plug as for the ways that were passing—ways that would not return in my time, or in anyone else's.[34]

Joshua Lee was correct. By the 1950s, the southern mule's fate no longer hung in the balance. Tractors, cotton pickers, chemicals, industrial jobs, and a host of other factors acted to destroy the underpinnings of an agricultural and social system which the mule had been part of for a century.

If few contemporaries mourned the end of the mule era, some individuals, such as Joshua Lee, clearly recognized the depth of the change in the South. Thus wrote David M. Potter in 1961, even as the mule South died,

> The enigma remains, and the historian must still ask what distinctive quality it was in the life of the South for which Southerners have felt such a persistent, haunting nostalgia and to which even the Yankee has responded with a poignant impulse. We must now doubt that this nostalgia was the yearning of men for an ideal agrarian utopia which never existed in reality. But if it was not that, was it perhaps the yearning of men in a mass culture for the life of a folk culture which did really exist? This folk culture, we know, was far from being ideal or utopian, and was in fact full of inequality and wrong, but if the nostalgia persists was it because even the inequality and wrong were parts of a life that still had a relatedness and meaning which our more bountiful life in the mass culture seems to lack?[35]

The southern folk culture to which Potter refers rested above all else upon relationships among living creatures, and partly upon the kinship between man and mule who battled the boll weevil, the elements, and chance to produce a new crop each year. Perhaps it was the loss of humanity in the face of the machine that pricked at many and forced them to consider a simpler past when skills such as judging good mule flesh were important. Throughout the mule era, to be considered a reliable judge of mules had broader applications in life. To know the difference between good and bad mule deals meant that one understood animals and the people who bought, sold, and traded them. Joshua Lee's eulogy to his grandfather, James Brantley Tolbert II, summed up much:

"To J.B., from those who knew him in the fullness of his powers, a candid solid man who drank deeply from life's cup, a man who regarded each boy as a son, and who found delight in every little girl, a born leader, a tiller of the soil, and a shrewd judge of mules."[36]

The story of the mule, then, is the story of a changing and changed South. For mule adherents, the tale is a sad one, replete with dislocation and death befitting a war. For others, such as Dennis S. Nordin and Roy V. Scott, the shift to agribusiness is a wholly positive tale because it has provided cheap food to a growing population.[37] But perhaps the story is not completely over. In the past twenty years or so, draft animals, including mules, have been revived across the nation as an alternative to modern agricultural techniques.

One of the most eloquent and widely read proponents of draft animals is farmer, poet, novelist, and critic Wendell Berry. Berry's books, such as *The Unsettling of America,* point at fault lines in American society and offer solutions to those ills. In terms of mechanization, Berry asserts that at some point, "*more* begins to imply *worse.* The mechanization of farming passed that point long ago—probably when it passed from horse power to tractor power."[38] Another critic of modern agriculture, Lynn Miller, founder of the *Small Farmer's Journal,* has for much of his life sounded a consistent call for a simpler, more fulfilling life. As they do for Berry, draft animals form an integral part of Miller's vision with the focus on small, diversified farms that add to community strength. Both Berry and Miller are, in David B. Danbom's words, "romantic agrarians" in the sense that they argue that contact with the land is central to a society's health, but that any contact with the soil is nonexistent for most and threatened for the rest.[39] Victor Davis Hanson has recently asserted that the nation is now beyond the stage of romantic agrarianism, however. As he writes, "We are now in the third stage of a future that has no future, an agrarian Armageddon at the millennium where the family farm itself—both as a way of life and a reassuring image of the mind—will be obliterated."[40]

There is a practical side to the agrarian vision, or at least thousands and thousands of horse farmers believe there is. The draft-horse industry bottomed out in the 1960s and 1970s, but breed associations are active once again. Numerous journals reflect a healthy draft-horse trade. Draft-horse clinics and schools are filled with individuals who want to learn about using the animals on farms. The interest in traditional modes of living and the number of draft-horse clinics are on the rise, and people from all walks of life attend the gatherings. One may find at these gatherings attorneys, farmers, foresters, nurses, and professors. For some individuals, draft animals are an interesting, but ultimately unimportant, throwback to an earlier era, but for others, draft horses and mules offer a means of breaking out of what they consider the traps of modern life.

The epic struggle between mules and machines thus ended with the triumph of urban, industrial values over tradition. Many elements of industrial society, from electric lights to radios, took their place on southern farms, but none altered the very nature of farming as did the tractor. Automobiles and trucks changed the way a farmer transported himself, his family, and his crops, but a tractor profoundly modified his relationship to the land and to the outside world because he by necessity became fully enmeshed in corporate capitalism. With the purchase of a tractor, the farmer became wholly dependent upon industrial giants such as International Harvester for the power to produce.

What does all of this have to do with the South? The South was late to mechanize, and much has been written concerning the possibilities for a different kind of agricultural South. The mule South is dead, although traditional mores and customs are still part of the region's memory and character, even if they are fading. The southern worldview continues to be more conservative than that of the nation as a whole in relation to political, religious, and moral issues, as well as the role of the family and women.[41] So vestiges remain, but these things too shall pass, to paraphrase my grandmother's way of comforting my usually nervous grandfather in times of stress. In the case of the American South, the real question is, as my grandfather replied, "But what the hell is next?" Time will tell.

Abbreviations

D&PL-AS	Delta and Pine Land Company, Delta and Pine Land Company Annual Statements and Presidents Reports, Special Collections, Mitchell Memorial Library, Mississippi State University
D&PL-OJ	Delta and Pine Land Company, Oscar Johnston, General Correspondence, 1933–39, Special Collections, Mitchell Memorial Library, Mississippi State University
ESAR	Extension Service Annual Report, Records of the Federal Extension Service, Record Group 33, National Archives
HAA	Horse Association of America
PHC	*The Papers of Henry Clay*
ROSA-GC	Records of the Office of the Secretary of Agriculture, General Correspondence of the Office of the Secretary, 1906–70, Record Group 16, National Archives
USCO	United States Census Office
USDA	United States Department of Agriculture
WGW	*The Writings of George Washington from the Original Manuscript Sources, 1745–1799*
TAS	*The American Slave: A Composite Autobiography*, ed. George P. Rawick, 1st supplement series, 12 vols. (Westport, CT: Greenwood Press, 1977)

Notes

Introduction

1. James C. Cobb, *Away down South: A History of Southern Identity* (New York: Oxford University Press, 2005), 138.

2. Jerry Leath Mills, "The Dead Mule Rides Again," *Southern Cultures* 6 (Winter 2000): 15. For an earlier version of his argument, see Mills, "Equine Gothic: The Dead Mule as Generic Signifier in Southern Literature of the Twentieth Century," *Southern Literary Journal* 29 (Fall 1996): 2–17.

3. The South, of course, is not the only place where mules have been important. See, for example, Hilaire Belloc, "Study of a Mule," *Commonweal* 13 (December 3, 1930): 127–28.

4. Margaret E. Derry, *Horses in Society: A Story of Animal Breeding and Marketing Culture, 1800–1920* (Buffalo: University of Toronto Press, 2006), xi–xii.

5. Ann Norton Greene, "Harnessing Power: Industrializing the Horse in Nineteenth Century America" (PhD diss., University of Pennsylvania, 2004), 4.

6. Roger Wade West, "'God's eunuch race': The Mule in Southern Literature and Folk Culture" (PhD diss., University of Southern Mississippi, 1996), 3.

7. James C. Cobb, *The Mississippi Delta and the World: The Memoirs of David L. Cohn* (Baton Rouge: Louisiana State University Press, 1995), 63.

8. Steven Stoll, *Larding the Lean Earth: Soil and Society in Nineteenth-Century America* (New York: Hill and Wang, 2002), 8.

9. S. Jonathan Bass, "'How 'bout a Hand for the Hog'": The Enduring Nature of the Swine as a Cultural Symbol in the South," *Southern Cultures* 1 (Spring 1995): 301–20.

10. Cobb, *Away down South* 7.

11. Emmett M. Essin, *Shavetails and Bell Sharps: The History of the U.S. Army Mule* (Lincoln: University of Nebraska Press, 1997), 69–88; John D. Billings, *Hardtack and Coffee: The Unwritten Story of Army Life* (Boston: G. M. Smith, 1887; reprint, Lincoln: University of Nebraska Press, 1993), 279–97; John Solomon Otto, *Southern Agriculture during*

the Civil War Era, 1860–1880 (Westport, CT: Greenwood Press, 1994), 19–45; and Derry, *Horses in Society.*

12. Harry L. Watson, "Front Porch," *Southern Cultures* 6 (Winter 2000): 5.

13. Wendell Berry, *The Gift of Good Land: Further Essays Cultural and Agricultural* (San Francisco: North Point Press, 1981), 111–12.

14. Stoll, *Larding the Lean Earth* 4.

15. Anthony Dent, *Donkey: The Story of the Ass from East to West* (London: George G. Harrap, 1972), 62–63.

16. Albert C. Leighton, "The Mule as a Cultural Invention," *Technology and Culture* 8 (January 1967): 47.

17. Kyle D. Kauffman, "Why Was the Mule Used in Southern Agriculture? Empirical Evidence of Principal-Agent Solutions," *Explorations in Economic History* 30 (July 1993): 338.

18. Leighton, "Mule as a Cultural Invention" 47.

19. Hilton M. Briggs and Dinus M. Briggs, *Modern Breeds of Livestock* (New York: Macmillan, 1980), 770–71.

20. J. O. Williams and William Jackson, "Improving Horses and Mules," *Yearbook of Agriculture, 1936* (Washington, D.C., 1936), 942.

21. Unless otherwise noted, the discussion of breeds is a summary of the information contained in "Breeds of Jacks," *American Jack Stock Stud Book* (Nashville: American Breeders Association of Jacks and Jennets, 1891), 1:36–56.

22. Briggs and Briggs, *Modern Breeds of Livestock* 772.

23. Dent, *Donkey* 112.

24. Summarized from a 1930s International Correspondence Schools instructional pamphlet entitled "Mules," reprinted in part in *Small Farmer's Journal* 21 (Winter 1998): 64–65.

25. "Mules," *Small Farmer's Journal* 63, 68.

26. John Langdon, *Horses, Oxen, and Technological Innovation: The Use of Draught Animals in English Farming from 1066 to 1500* (Cambridge: Cambridge University Press, 1986), 8.

27. Dent, *Donkey* 67–68; Harold B. Barclay, *The Role of the Horse in Man's Culture* (New York: J. A. Allen, 1980), 125.

28. Barclay, *Role of the Horse* 131–32.

29. Leighton, "Mule as a Cultural Invention" 52.

30. William Faulkner, *Absalom, Absalom!* (New York: Random House, 1936), 162.

Chapter 1

1. USCO, *Twelfth Census of the United States, 1900, Agriculture* (Washington, D.C., 1902), cxcv.

2. For a graphical treatment of this point, see Sam Bowers Hilliard, *Atlas of Antebellum Southern Agriculture* (Baton Rouge: Louisiana State University Press, 1984), 47.

3. Greene, "Harnessing Power" 90.

4. Wilbur Cash, *The Mind of the South* (New York: Alfred A. Knopf, 1941), 9–10.

5. See William M. Mathew, *Edmund Ruffin and the Crisis of Slavery in the Old South: The Failure of Agricultural Reform* (Athens: University of Georgia Press, 1988); Stoll, *Larding the Lean Earth.*

6. William J. Cooper, "The Cotton Crisis in the Antebellum South: Another Look," *Agricultural History* 49 (April 1975): 381–91.

7. Alan I. Marcus and Howard P. Segal, *Technology in America: A Brief History* (New York: Harcourt Brace Jovanovich, 1989), 96, 98–101.

8. John Niven, *The Coming of the Civil War, 1837–1861* (Arlington Heights, IL: Harlan Davidson, 1990), 9.

9. John McCardell, *The Idea of a Southern Nation: Southern Nationalists and Southern Nationalism, 1830–1860* (New York: W. W. Norton, 1979), 3–9.

10. Mark M. Smith, *Listening to Nineteenth-Century America* (Chapel Hill: University of North Carolina Press, 2001), 17.

11. Elizabeth Fox-Genovese, "The Anxiety of History: The Southern Confrontation with Modernity," *Southern Cultures* Inaugural Issue (1993): 65–82.

12. *Southern Planter* (May 1860): 274. For a full treatment of the role of time in southern thought, see Mark M. Smith, *Mastered by the Clock: Time, Slavery, and Freedom in the American South* (Chapel Hill: University of North Carolina Press, 1997).

13. *Southern Planter* (January 1852): 12.

14. Joseph P. Reidy, *From Slavery to Agrarian Capitalism in the Cotton Plantation South: Central Georgia, 1800–1880* (Chapel Hill: University of North Carolina Press, 1992), 39–40.

15. *Southern Planter* (June 1841): 92.

16. *Southern Planter* (January 1841): 10.

17. Smith, *Mastered by the Clock* 4.

18. Cornelius Oliver Cathey, *Agricultural Developments in North Carolina, 1783–1860* (Chapel Hill: University of North Carolina Press, 1956), 35, 184.

19. John Hebron Moore, *The Emergence of the Cotton Kingdom in the Old Southwest: Mississippi, 1770–1860* (Baton Rouge: Louisiana State University Press, 1988), 51–52.

20. J. Carlyle Sitterson, *Sugar Country: The Cane Sugar Industry in the South, 1753–1950* (Lexington: University of Kentucky Press, 1953), 52.

21. James C. Bonner, *A History of Georgia Agriculture, 1732–1860* (Athens: University of Georgia Press, 1964), 137.

22. USCO, *Eighth Census of the United States, 1860, Agriculture* (Washington, D.C., 1864), 22–27.

23. Bonner, *History of Georgia Agriculture* 137.

24. *Eighth Census* cx, cxii, cxiv. This census defined southern states as Alabama, Arkansas, Florida, Georgia, Mississippi, Louisiana, North Carolina, South Carolina, Tennessee, Texas, and Virginia.

25. *Eighth Census* cx, cxii, cxiv.

26. Stoll, *Larding the Lean Earth* 30.

27. Joyce Chaplin, *An Anxious Pursuit: Agricultural Innovation and Modernity in the Lower South, 1730–1815* (Chapel Hill: University of North Carolina Press, 1993), 252.

28. *American Farmer* (December 17, 1819): 299.

29. George F. Lemmer, "Early Agricultural Editors and Their Farm Philosophies," *Agricultural History* 31 (October 1957): 3–22.

30. Cooper, "Cotton Crisis in the Antebellum South" 383.

31. Lemmer, "Early Agricultural Editors" 3, 12.

32. Stoll, *Larding the Lean Earth* 26–27.

33. *Southern Cultivator* (June 1855): 180.

34. *Southern Cultivator* (November 1843): 189.

35. *Southern Cultivator* (February 1850): 21.

36. *American Farmer* (October 20, 1820): 239.

37. *Southern Planter* (May 1855): 129.

38. Ibid.

39. *Southern Planter* (January 1856): 13.

40. *Farmers' Cabinet* (October 15, 1842): 77.

41. *Southern Cultivator* (April 1851): 54.

42. *Southern Cultivator* (July 1843): 117–18.

43. *Farmers' Cabinet* (October 15, 1842): 77.

44. Bonner, *History of Georgia Agriculture* 56.

45. *Southern Cultivator* (November 1852): 323.

46. *Southern Planter* (November 1855): 351.

47. *American Farmer* (December 1819): 299.

48. Greene, "Harnessing Power" 75.

49. *American Farmer* (December 1819): 299.

50. *Southern Planter* (January 1844): 4.

51. *Southern Cultivator* (February 1850): 21.

52. *American Farmer* (December 1819): 299.

53. Richard L. Troutman, "Plantation Life in the Ante-bellum Bluegrass Region of Kentucky" (master's thesis, University of Kentucky, 1955), 77; *Southern Cultivator* (June 1850): 84.

54. *Southern Cultivator* (August 1849): 116.

55. *American Farmer* (December 1819): 299.

56. Robert Byron Lamb, *The Mule in Southern Agriculture* (Berkeley: University of California Press, 1963), 35.

57. Horse and Mule Association of America, *Jacks, Jennets, and Mules* (Chicago: HAA, 1945), 25.

58. *American Farmer* (December 1819): 299.

59. James O. Farmer Jr., *The Metaphysical Confederacy: James Henley Thornwell and the Synthesis of Southern Values* (Macon, GA: Mercer University Press, 1986), 5.

60. *Southern Cultivator* (July 19, 1843): 117.

61. Percy Wells Bidwell and John I. Falconer, *History of Agriculture in the Northern United States, 1620–1860* (Washington, D.C.: Carnegie Institution of Washington, 1925), 111.

62. *American Farmer* (December 8, 1821): 291.

63. Bidwell and Falconer, *History of Agriculture* 113.

64. Report of the Commissioner of Patents for the Year 1854, Agriculture (Washington, D.C., 1855), 38.

65. Ibid., 42; Report of the Commissioner of Patents for the Year 1853, Agriculture (Washington, D.C., 1854), 34.

66. *Farmers' Cabinet* (October 15, 1842): 78.

67. *Eighth Census* cxii, cx, and cxiv.

68. The New England region included Connecticut, Maine, Massachusetts, New Hampshire, Rhode Island, and Vermont (*Eighth Census* cxi).

69. *American Agriculturist* (July 1846): 219.

70. *Southern Planter* (November 1858): 693.

71. Harvey Riley, *The Mule: A Treatise on the Breeding, Training, and Uses to Which He May Be Put* (New York: Dick and Fitzgerald, 1867), 3, 36–38.

72. *Farmers' Cabinet* (October 15, 1842): 77.

73. *Southern Planter* (April 1861): 228.

74. *De Bow's Review* (April 1850): 388.

75. *Southern Planter* (November 1858): 693.

76. J. S. Skinner, "The Ass and the Mule," in William Youatt, *The Horse* (New York: Leavitt and Allen, 1843), 431; *Southern Planter* (January 1856): 13.

77. *Southern Planter* (March 1850): 84.

78. *De Bow's Review* (April 1850): 388.

79. *American Farmer* (March 1821): 408.

80. *Southern Planter* (March 1850): 84.

81. *De Bow's Review* (April 1850): 388.

82. François André Michaux, *Travels to the Westward of the Allegany Mountains, in the States of the Ohio, Kentucky, and Tennessee, and Return to Charlestown, through the Upper Carolinas* (London: J. Mawman, 1805), 66, 233–34.

83. G. Terry Sharrer, *A Kind of Fate: Agricultural Change in Virginia, 1861–1920* (Ames: Iowa State University Press, 2000), especially chapter 1.

84. *American Agriculturist* (June 1846): 187; Michaux, *Travels* 233.

85. *American Agriculturist* (July 1846): 219.

86. *American Agriculturist* (June 1846): 187.

87. Jack Temple Kirby, *Rural Worlds Lost: The American South, 1920–1960* (Baton Rouge: Louisiana State University Press, 1987), 200.

88. *American Agriculturist* (November 1846): 339.

89. "James Bolton Interview," in *TAS, Georgia Narratives*, 3:85.

90. Ray Mathis, Mary Mathis, and Douglas Clare Purcell, eds., *John Horry Dent Farm Journals and Account Books, 1840–1892* (Tuscaloosa: published for the Historic Chatta-

hoochee Commission by University of Alabama Press, 1977), microfilm, May 17–18, 1856, 3:191. For a description of Dent and a complete index to his journals, see Ray Mathis and Mary Mathis, *Introduction and Index to the John Horry Dent Farm Journals and Account Books, 1840–1892* (Tuscaloosa: University of Alabama Press, 1977).

91. Skinner, "The Ass and the Mule" 431.

92. *De Bow's Review* (April 1850): 387.

93. *Southern Planter* (April 1861): 228.

94. Kyle Dean Kauffman, "The Use of Draft Animals in America: Economic Factors in the Choice of an Early Motive Power" (PhD diss., University of Illinois at Urbana-Champaign, 1993), 4–5.

95. *Southern Cultivator* (April 26, 1843): 62.

96. *Southern Planter* (January 1852): 12.

97. *De Bow's Review* (April 1850): 388.

98. Report of the Commissioner of Patents for the Year 1853, Agriculture, 34.

99. *Southern Cultivator* (April 1851): 54.

100. Mathis, Mathis, and Purcell, *Dent Farm Journals*, February 26, 1856, 3:172.

101. Mathis, Mathis, and Purcell, *Dent Farm Journals*, February 8, 1859, 5:37.

102. Mathis, Mathis, and Purcell, *Dent Farm Journals*, February 21, 1860, 5:151.

103. Bonner, *History of Georgia Agriculture* 138.

104. *Southern Planter* (April 1861): 228.

105. Kauffman, *Use of Draft Animals* 6, 9 (fig. 1.1, "Real Prices of Mules and Horses").

106. *Southern Cultivator* (April 26, 1843): 62.

107. John Frederic Olson, "The Occupational Structure of Plantation Slave Labor in the Late Antebellum Era" (PhD diss., Rochester, NY: University of Rochester, 1983), 67–68.

108. Ibid., 71.

109. Mathis, Mathis, and Purcell, *Dent Farm Journals*, February 5–23, 5:14–22.

110. Acquisition of Property, Other Property: Inventories of Property 1852–1866 and Undated; Rice Collection, Nannie Herndon Rice Family Papers, John Washington Rice Estate, Mitchell Memorial Library, Mississippi State University.

111. Olson, "Occupational Structure" 71.

112. See Grady McWhiney, introduction to Frank L. Owsley, *Plain Folk of the Old South* (Baton Rouge: Louisiana State University Press, 1982); Forrest McDonald and Grady McWhiney, "The Antebellum Southern Herdsman: A Reinterpretation," *Journal of Southern History* 41 (May 1975): 147–66; and Stephanie McCurry, *Masters of Small Worlds: Yeoman Households, Gender Relations, and the Political Culture of the Antebellum South Carolina Low Country* (New York: Oxford University Press, 1995).

113. See Steven Hahn, *The Roots of Southern Populism: Yeoman Farmers and the Transformation of the Georgia Upcountry, 1850–1890* (New York: Oxford University Press, 1983); Lacy K. Ford, *Origins of Southern Radicalism: The South Carolina Upcountry, 1800–1860* (New York: Oxford University Press, 1988); James Oakes, *Slavery and Freedom: An*

Interpretation of the Old South (New York: Alfred A. Knopf, 1990); and McCurry, *Masters of Small Worlds.*

114. See Harold D. Woodman, *King Cotton and His Retainers: Financing and Marketing the Cotton Crop of the South, 1800–1925* (Lexington: University of Kentucky Press, 1968); and Gavin Wright, *The Political Economy of the Cotton South: Households, Markets, and Wealth in the Nineteenth Century* (New York: W. W. Norton, 1978).

115. Fred A. Bailey, "Class and Tennessee's Confederate Generation," *Journal of Southern History* 51 (February 1985): 60. See also James Oakes, *The Ruling Race: A History of American Slaveholders* (New York: Alfred A. Knopf, 1982); and Oakes, "From Republicanism to Liberalism: Ideological Change and the Crisis of the Old South," *American Quarterly* 37 (Fall 1985): 551–71.

116. Steven Hahn, "The Yeomanry of the Nonplantation South: Upper Piedmont Georgia, 1850–1860," in *Class, Conflict, and Consensus: Antebellum Southern Community Studies,* ed. Orville Vernon Burton and Robert C. McMath Jr. (Westport, CT: Greenwood Press, 1982), 36.

117. *Southern Planter* (August 1855): 251.

118. *Eighth Census,* 154–55.

119. Terry G. Jordan, "The Imprint of the Upper and Lower South on Mid-Nineteenth Century Texas," *Annals of the Association of American Geographers* 57 (December 1967): 667–90.

120. *Southern Planter* (January 1852): 12.

121. Lamb, *Mule in Southern Agriculture* 63, 65.

122. Ellis Merton Coulter, *Old Petersburg and the Broad River Valley of Georgia: Their Rise and Decline* (Athens: University of Georgia Press, 1965), 23; and *Eighth Census,* 22.

123. Ford, *Origins of Southern Radicalism* 46.

124. USCO, *Seventh Census of the United States, 1850* (Washington, D.C., 1853), 345; and *Eighth Census,* 128.

125. Ford, *Origins of Southern Radicalism* 46.

126. *Eighth Census,* 128.

127. Ibid.

128. For a full discussion of Georgia's economic regions, see Frederick A. Bode and Donald E. Ginter, *Farm Tenancy and the Census in Antebellum Georgia* (Athens: University of Georgia Press, 1986), especially chapter 4.

129. Steven Hahn, "The 'Unmaking' of the Southern Yeomanry: The Transformation of the Georgia Upcountry, 1860–1890," in *The Countryside in the Age of Capitalist Transformation: Essays in the Social History of Rural America,* ed. Steven Hahn and Jonathan Prude (Chapel Hill: University of North Carolina Press, 1985), 183; and Eugene D. Genovese, "Yeomen Farmers in a Slaveholders' Democracy," *Agricultural History* 49 (April 1975): 331–42.

130. Wright, *Political Economy of the Cotton South* 55.

131. Randolph B. Campbell, "Intermittent Slave Ownership: Texas as a Test Case," *Journal of Southern History* 51 (February 1985): 15–23; James Oakes, "['Intermittent Slave Ownership: Texas as a Test Case']: A Response," *Journal of Southern History* 51 (February 1985): 23–28; and Randolph B. Campbell, "['Intermittent Slave Ownership: Texas as a Test Case']: A Rejoinder," *Journal of Southern History* 51 (February 1985): 29–30.

132. Lamb, *Mule in Southern Agriculture* 34 (table 3, "Mules in the South, 1860–1900").

133. Ibid., 37.

134. Ibid., 37–43.

135. Ibid., 45.

136. W. Hustace Hubbard, *Cotton and the Cotton Market* (New York: D. Appleton, 1923), 21.

Chapter 2

1. Leighton, "Mule as a Cultural Invention" 50.

2. Alfred Glaze Smith Jr., *Economic Readjustment of an Old Cotton State: South Carolina, 1820–1860* (Columbia: University of South Carolina Press, 1958), 79–80.

3. Michaux, *Travels* 232–33.

4. Riley, *Mule* 34; Cathey, *Agricultural Developments* 186; and *Southern Cultivator* (April 26, 1843): 62.

5. HAA, *Jacks, Jennets, and Mules* 4.

6. Dent, *Donkey* 105–6.

7. *Southern Planter* (July 1853): 237.

8. Lamb, *Mule in Southern Agriculture* 32.

9. Lewis Cecil Gray, *History of Agriculture in the Southern United States to 1860* (New York: Peter Smith, 1941), 80.

10. Dent, *Donkey* 111–12.

11. Edwin Morris Betts, *Thomas Jefferson's Farm Book* (Charlottesville: University Press of Virginia, 1976), 48, 110.

12. Gray, *History of Agriculture* 612.

13. *Southern Planter* (July 1853): 237.

14. George Washington to Robert Townsend Hooe, July 18, 1784, in *WGW*, ed. John C. Fitzpatrick (Washington, D.C., 1938), 27:447.

15. George Washington to Marquis de Lafayette, September 1, 1785, in *WGW*, 27:447.

16. George Washington to Robert Townsend Hooe, July 18, 1784, in *WGW*, 27:447.

17. Much of the information in these paragraphs comes from the detailed article by J. H. Powell, "General Washington and the Jack Ass," *South Atlantic Quarterly* 52 (April 1953): 238–52.

18. HAA, *Jacks, Jennets, and Mules* 4; Kirby, *Rural Worlds Lost* 196.

19. Worth Estes, "Henry Clay as a Livestock Breeder," *Filson Club History Quarterly* 32 (October 1958): 352.

20. Amelia Clay Van Meter Rogers, "Ashland: The Home of Henry Clay" (master's thesis, University of Kentucky, 1934), 19.

21. Richard Laverne Troutman, "Henry Clay and His 'Ashland' Estate," *Filson Club History Quarterly* 30 (April 1956): 167.

22. Rogers, "Ashland" 16.

23. Henry Clay to Richard B. Jones, July 5, 1831, in *PHC*, vol. 8, ed. Robert Seager II and Melba Porter Hay (Lexington: University Press of Kentucky, 1984), 371.

24. Troutman, "Henry Clay" 168.

25. John H. Kerr to Henry Clay, October 15, 1827, in *PHC*, vol. 6, ed. Mary W. M. Hargreaves and James F. Hopkins (Lexington: University Press of Kentucky, 1981), 1151.

26. Advertisement, June 24–25, 1825, in *PHC*, vol. 4, ed. James F. Hopkins, Mary W. M. Hargreaves et al. (Lexington: University Press of Kentucky, 1972), 435.

27. Richard L. Troutman, "The Social and Economic Structure of Kentucky Agriculture, 1850–1860" (PhD diss., University of Kentucky, 1958), 96.

28. Henry Clay to Richard B. Jones, July 5, 1831, in *PHC* 8:371.

29. Troutman, "Henry Clay" 168.

30. Gray, *History of Agriculture* 854.

31. Henry Clay to S. Bennett, June 23, 1837, in *PHC*, vol. 9, ed. Robert Seager II and Melba Porter Hay (Lexington: University Press of Kentucky, 1988), 53.

32. Troutman, "Henry Clay" 169.

33. Troutman, "Plantation Life" 74; Bidwell and Falconer, *History of Agriculture* 447.

34. Troutman, "Henry Clay" 168; Agreement with John Rodgers, November 20, 1827, in *PHC* 6:1296.

35. John Rodgers to Henry Clay, March 13, 1829, in *PHC* 8:34.

36. Henry Clay to Richard B. Jones, July 5, 1831, in *PHC* 8:371.

37. Troutman, "Henry Clay" 168.

38. Estes, "Henry Clay as a Livestock Breeder" 352.

39. Troutman, "Henry Clay" 169; Gray, *History of Agriculture* 852.

40. Henry Clay to Thomas Hart Clay, November 25, 1833, in *PHC* 8:668.

41. Henry Clay to Henry Clay Jr., May 1, 1832, in *PHC* 8:502.

42. Henry Clay Jr. to Henry Clay, May 11, 1832, in *PHC* 8:509.

43. Troutman, "Henry Clay" 168.

44. L. W. Knight, "Reminiscences of Jacks and Jack Breeders," in *American Jack Stock Stud Book* (Nashville: American Breeders Association of Jacks and Jennets, 1891), 1:26–27.

45. Briggs and Briggs, *Modern Breeds of Livestock* 773.

46. Rezin D. Shepherd to Henry Clay, March 13, 1836, in *PHC* 8:835.

47. Troutman, "Henry Clay" 169.

48. J. T. Warder, "Mule Raising," in *Annual Report of the Commissioner of Agriculture, 1863* (Washington, D.C., 1863), 185.

49. Henry Clay to James Erwin, October 13, 1833, in *PHC* 8:665; Henry Clay to Nathaniel Hart, January 30, 1836, in *PHC* 8:826.

50. *Southern Cultivator* (November 1847): 172.

51. Troutman, "Plantation Life" 75.

52. *Southern Planter* (October 1854): 297.

53. *Southern Cultivator* (April 26, 1843): 62.

54. Gray, *History of Agriculture* 854.

55. *Southern Planter* (October 1854): 297–98. The name was also spelled "Cockrill" in some articles.

56. Henry Clay to S. Bennett, June 23, 1837, in *PHC* 9:53.

57. Terry G. Jordan, "The Imprint of the Upper and Lower South on Mid-Nineteenth Century Texas," *Annals of the Association of American Geographers* 57 (December 1967): 669.

58. John Solomon Otto, *Southern Agriculture during the Civil War Era, 1860–1880* (Westport, CT: Greenwood Press, 1994), 93.

59. Michaux, *Travels* 234.

60. *Camden [South Carolina] Journal* as quoted in *Nile's Register*, in Elizabeth L. Parr, "Kentucky's Overland Trade with the Ante-Bellum South," *Filson Club History Quarterly* 2 (January 1928): 76.

61. Parr, "Kentucky's Overland Trade" 75.

62. Mary Verhoeff, *The Kentucky Mountains*, Filson Club Publications No. 25 (Louisville: Filson Club, 1911), 146; Parr, "Kentucky's Overland Trade" 72–73, 75.

63. *Southern Planter* (April 1894): 201.

64. "100 Years Ago Today: 1906," *[Memphis] Commercial Appeal* May 21, 2006, B2.

65. Robert Louis Jones, "The Horse and Mule Industry in Ohio to 1865," *Mississippi Valley Historical Review* 33 (June 1946): 86; Troutman, "Plantation Life" 78.

66. *Southern Planter* (November 1858): 693.

67. Gray, *History of Agriculture* 840; Parr, "Kentucky's Overland Trade" 77.

68. Parr, "Kentucky's Overland Trade" 74; Michaux, *Travels* 234.

69. Troutman, "Social and Economic Structure" 103.

70. *Southern Planter* (November 1858): 692–93.

71. *Southern Planter* (October 1872): 592.

72. Troutman, "Plantation Life" 76.

73. Bidwell and Falconer, *History of Agriculture* 447.

74. George H. Dacy, "Missouri Mules Make Money," *Breeder's Gazette* 78 (August 26, 1920): 335.

75. Parr, "Kentucky's Overland Trade" 73–74; Thomas D. Clark, "The Trade between Kentucky and the Cotton Kingdom in Livestock, Hemp and Slaves from 1840 to 1860" (master's thesis, University of Kentucky, 1929), 37; *Southern Planter* (June 1868): 335–37.

76. Dacy, "Missouri Mules Make Money" 335.

77. Paul C. Heinlein, "Shifting Range-Feeder Patterns in the Ohio Valley before 1860," *Agricultural History* 31 (January 1957): 1.

78. *Southern Planter* (November 1858): 692.

79. Report of the Commissioner of Patents for the Year 1853, Agriculture, 29–30.

80. "History of Organization," in *American Jack Stock Stud Book* (Nashville: American Breeders Association of Jacks and Jennets, 1891), 1:11–12.

81. Briggs and Briggs, *Modern Breeds of Livestock* 773.

82. "History of Organization" 1:11–12.

83. J. J. Hooper and W. S. Anderson, "Jack Stock of Kentucky," in *Kentucky Agricultural Experiment Station Bulletin* 176 (November 30, 1913), 362.

84. For a summary of mule importations during the 1880s, see Knight, "Reminiscences of Jacks and Jack Breeders" 1:30–35.

85. Harry Bernstein, "Spanish Influence in the United States: Economic Aspects," *Hispanic American Historical Review* 18 (February 1938): 49.

86. Bidwell and Falconer, *History of Agriculture* 447.

87. Dacy, "Missouri Mules Make Money" 335.

88. C. Allan Jones, *Texas Roots: Agriculture and Rural Life before the Civil War* (College Station: Texas A&M University Press, 2005), 192–94.

89. Odie B. Faulk, "Ranching in Spanish Texas," *Hispanic American Historical Review* 45 (May 1965): 265.

90. See William R. Ferris, *Mule Trader: Ray Lum's Tales of Horses, Mules, and Men* (Jackson: University Press of Mississippi, 1998).

91. Kemp Hill, "Horse Trader," interview by Harry H. Fain, March 20, 1939, typescript, Federal Writers' Project Life History Interviews, Southern Historical Collection, Wilson Library, University of North Carolina.

92. Joshua A. Lee, *With Their Ears Pricked Forward: Tales of Mules I've Known* (Winston-Salem, NC: John F. Blair, 1980), 1.

93. L. W. Knight, *The Breeding and Rearing of Jacks, Jennets and Mules* (Nashville: Cumberland Press, 1902), 13.

94. "Keith Jones Interview," October 13, 1982, in *Recollections of Missouri Mules*, ed. Melvin Bradley and Duane Dailey (Columbia: University of Missouri–Columbia, 1991), 4:2.

95. Oscar Johnston to J. W. McCluer, September 21, 1932, D&PL-OJ.

96. Lawrence J. Nelson, "Welfare Capitalism on a Mississippi Plantation in the Great Depression," *Journal of Southern History* 50 (May 1984): 230.

97. Oscar Johnston to J. W. McCluer, October 12, 1932, D&PL-OJ.

98. Nelson, "Welfare Capitalism" 230; Read P. Dunn Jr., *Mr. Oscar: A Story of the Early Years in the Life and Times of Oscar Johnston and of His Efforts in Organizing the National Cotton Council* (Memphis, TN: National Cotton Council of America, 1991), 16.

99. Oscar Johnston to D. S. Lantrip, November 15, 1937, D&PL-OJ.

100. D&PL-AS (1926), 74.

101. Oscar Johnston to Dr. C. McMillan, July 12, 1939, D&PL-OJ.

102. Oscar Johnston to Unit Managers, December 9, 1940, D&PL-OJ; D&PL-AS (1946), handwritten ledger sheet; Oscar Johnston to Unit Managers, January 16, 1942, D&PL-OJ; and Oscar Johnston to J. W. McCluer, December 8, 1941, D&PL-OJ.

103. Oscar Johnston to J. W. McCluer, December 7, 1940, D&PL-OJ.

104. Pete Daniel, *Breaking the Land: The Transformation of Cotton, Tobacco, and Rice Cultures since 1880* (Urbana: University of Illinois Press, 1985), 170.

105. D&PL-AS (1940), 3–4; D&PL-AS (1941), 3–4.

106. Nelson, "Welfare Capitalism" 244–45; Daniel, *Breaking the Land* 100–105; and James C. Cobb, *The Most Southern Place on Earth: The Mississippi Delta and the Roots of Regional Identity* (New York: Oxford University Press, 1992), 189–92.

107. Hill, "Horse Trader."

108. Lee, *With Their Ears Pricked Forward* 2.

109. Hill, "Horse Trader."

110. Ibid.

111. Ibid.

112. Roger Butterfield, "Missouri Mule Trader," *Life* (January 26, 1948): 111.

113. Ibid.

114. Ibid., 105–6, 108, 111–12, 114.

115. "Keith Jones Interview" 4:13.

116. See M. J. Danner, B. H. Luebke, and B. D. Raskopf, *Facilities and Agencies at Nashville Livestock Market*, University of Tennessee Agricultural Experiment Station Rural Research Series Monograph 206 (Knoxville, August 6, 1946).

117. Ibid., 27.

118. Harry Crews, "The Mythic Mule," *Southern Magazine* 1 (January 1987): 22.

119. Butterfield, "Missouri Mule Trader" 108, 111.

120. Lee, *With Their Ears Pricked Forward* 2–3.

121. Butterfield, "Missouri Mule Trader" 114.

122. Frank C. Mills, *History of American Jacks and Mules* (Hutchinson, KS: Hutch-Line, 1971), 70.

123. Knight, *Breeding and Rearing of Jacks, Jennets and Mules* 49.

124. *Southern Planter* (November 1858): 692.

125. Bonner, *History of Georgia Agriculture* 138.

126. Mills, *History of American Jacks and Mules* 70.

127. "Mule Mixup," *Time* 50 (July 14, 1947): 87.

128. Danner, Luebke, and Raskopf, *Facilities and Agencies* 27.

129. Hill, "Horse Trader."

130. "Mule Mixup," 86–87; Butterfield, "Missouri Mule Trader" 105–6.

Chapter 3

1. *Southern Cultivator* (June 1850): 84.

2. Bonner, *History of Georgia Agriculture* 138; Cathey, *Agricultural Developments* 185; Moore, *Emergence of the Cotton Kingdom* 55; and Owsley, *Plain Folk* 48.

3. *Southern Cultivator* (July 1855): 206.

4. Paul W. Gates, *The Farmer's Age: Agriculture, 1815–1860* (New York: Harper Torchbooks, 1960), 230.

5. *Southern Cultivator* (November 1847): 172.

6. Report of the Commissioner of Patents for the Year 1850, Agriculture (Washington, D.C., 1851), 188.

7. *Southern Planter* (May 1860): 274.

8. *Southern Cultivator* (June 1850): 84.

9. *Southern Cultivator* (November 1847): 172.

10. Report of the Commissioner of Patents for the Year 1851, Agriculture (Washington, D.C., 1852), 320.

11. Ibid., 331.

12. *Southern Cultivator* (June 1850): 84.

13. Julius Rubin, "The Limits of Agricultural Progress in the Nineteenth-Century South," *Agricultural History* 49 (April 1975): 362–73.

14. *Southern Cultivator* (March 1856): 91.

15. *Southern Cultivator* (July 1855): 206.

16. *Southern Cultivator* (March 1856): 91.

17. Report of the Commissioner of Patents for the Year 1849, Agriculture (Washington, D.C., 1850), 312.

18. *Southern Cultivator* (June 1850): 84.

19. Report of the Commissioner of Patents for the Year 1851, Agriculture, 335.

20. Ibid., 338.

21. Report of the Commissioner of Patents for the Year 1849, Agriculture, 310–11.

22. *Southern Planter* (March 1850): 84.

23. See Weymouth T. Jordan, *Hugh Davis and His Alabama Plantation* (Tuscaloosa: University of Alabama Press, 1948).

24. Gray, *History of Agriculture* 699.

25. Charles S. Davis, *The Cotton Kingdom in Alabama* (Montgomery: Alabama State Department of Archives and History, 1939), 186–87.

26. Report of the Commissioner of Patents for the Year 1853, Agriculture (Washington, D.C., 1854), 34.

27. Rubin, "Limits of Agricultural Progress" 362–73.

28. Warder, "Mule Raising" 183.

29. George Washington to William Pearce, July 13, 1794, in *WGW*, ed. John C. Fitzpatrick (Washington, D.C., 1940), 33:428.

30. J. O. Williams, *Mule Production*, United States Department of Agriculture Farmers' Bulletin No. 1341 (Washington, D.C., 1948), 6.

31. George Washington to Marquis de Lafayette, May 10, 1786, in *WGW*, ed. John C. Fitzpatrick (Washington, D.C., 1938), 28:423.

32. J. H. Powell, "General Washington and the Jack Ass," *South Atlantic Quarterly* 52 (April 1953): 246.

33. George Washington to John Hoomes, February 17, 1791, in *WGW*, ed. John C. Fitzpatrick (Washington, D.C., 1939), 31:218–19.

34. J. O. Williams, *Mule Production* 6.

35. *Southern Planter* (October 1854): 298.

36. *Southern Cultivator* (April 26, 1843): 62.

37. Riley, *Mule* 65.

38. *Southern Cultivator* (April 26, 1843): 62.

39. J. O. Williams, *Mule Production* 7.

40. A. L. French, "Raising Colts of Quality: The Day of the Near-Horse Is Passing," *Progressive Farmer* 35 (January 24, 1920): 23.

41. Wayne Dinsmore to Dr. R. A. Pearson, February 9, 1918, Box 512, ROSA-GC.

42. "Last Chance to Register Jacks and Jennets as Foundation Stock," *Progressive Farmer* 35 (May 15, 1920): 34.

43. Crews, "Mythic Mule" 23.

44. Report of the Commissioner of Patents for the Year 1853, Agriculture, 28.

45. Gates, *Farmer's Age* 230.

46. "Can the South Raise Its Own Horses?" *Southern Planter* (February 1873): 78.

47. *Southern Cultivator* (December 1874): 476.

48. Clarence Leroy Holmes, "Horses and Mules in American Agriculture, with Special Reference to Their Function in Agricultural Production" (PhD diss., University of Wisconsin, 1921), 59–60.

49. Hubbard, *Cotton* 22.

50. "Mules of the Right Type and How to Get Them," *Southern Agriculturist* 55 (March 1, 1925): 2.

51. Tait Butler, "What Are We Going to Do about Our Supply of Horses and Mules?" *Progressive Farmer* 41 (July 17, 1926): 772; Tait Butler, "Horse and Mule Shortage in Five Years Probable," *Progressive Farmer* 42 (June 25, 1927): 712.

52. Butterfield, "Missouri Mule Trader" 114.

53. Alfred Charles True, *A History of Agricultural Extension Work in the United States, 1785–1923*, USDA Miscellaneous Publication 15 (Washington, D.C., October 1928), 115.

54. ESAR, Macon County, Georgia, 1915; ESAR, Taliaferro County, Georgia, 1915; ESAR, Troup County, Georgia, 1915; and ESAR, Negro Agent Report, Amite County, Mississippi, 1915.

55. ESAR, Jackson County, Mississippi, 1935; ESAR, Tallahatchie County, Mississippi, 1935.

56. See Joseph Cannon Bailey, *Seaman A. Knapp: Schoolmaster of American Agriculture* (New York: Columbia University Press, 1975).

57. Memo from Rommel to Assistant Secretary Pearson, September 19, 1917, Box 346, ROSA-GC.

58. ESAR, Campbell County, Georgia, 1915; ESAR, Hancock County, Georgia, 1920.

59. ESAR, Bibb County, Georgia, 1915; ESAR, Dekalb County, Georgia, 1915.

60. ESAR, Jasper County, South Carolina, 1940.

61. USCO, Fifteenth Census of the United States, 1930, Agriculture (Washington, D.C., 1931–32), 58–59.

62. ESAR, Mississippi State Report-Animal Husbandry, 1936, 6; Ellis McFarland to *Southern Agriculturist* 47 (August 15, 1917): 12.

63. *Progressive Farmer* 35 (December 18, 1920): 200.

64. Tait Butler, "How Should Idle Work Stock Be Wintered?" *Progressive Farmer* 39 (January 12, 1924): 30.

65. G. E. Whitworth to D. F. Houston, September 30, 1919, Box 663, ROSA-GC; emphasis added.

66. USDA, "See Dearth of Farm Work Animals," *Official Record* 4 (May 6, 1925): 3.

67. USDA, "Shortage of Good Work Stock Expected in Five Years," *Official Record* 6 (September 14, 1927): 8.

68. USDA, "Improvement Noted: Horse and Mule Prices," *Official Record* 7 (April 18, 1928): 5.

69. USDA, *Official Record* 8 (August 22, 1929): 4.

70. USDA, "Horse and Mule Prices Hold Up," *Official Record* 11 (September 10, 1932): 193.

71. *Agricultural Outlook for South Carolina, 1931*, Clemson Agricultural College Extension Service, Circular 108 (Clemson, South Carolina, December 1930), 5.

72. ESAR, District Supervisor Report, Southwest District, Mississippi, 1935, 34.

73. *Southern Agriculturist* 65 (January 1935): 3.

74. ESAR, Subject Specialist Report-Animal Husbandry, South Carolina, 1941, 48.

75. Eugene D. Rutland, "Let's Raise Spring Colts," *Southern Agriculturist* 70 (February 1940): 34.

76. Herman Clarence Nixon, *Forty Acres and Steel Mules* (Chapel Hill: University of North Carolina Press, 1938), n.p.

77. Jack Wooten, "Raise Their Own," *Southern Agriculturist* 70 (January 1940): 19.

78. ESAR, Covington County, Mississippi, 1935, 14.

79. Wooten, "Raise Their Own" 19.

80. ESAR, Subject Specialist Report-Animal Husbandry, Mississippi, 1941, 48.

81. ESAR, District Supervisor Agent's Report, Southwest District, Mississippi, 1935, 34.

82. ESAR, Itawamba County, Mississippi, 1935, 13; ESAR, Smith County, Mississippi, 1935, 6.

83. ESAR, Abbeville County, South Carolina, 1935, 14.

84. *Southern Agriculturist* 65 (September 1935): 26.

85. *Southern Agriculturist* 65 (March 1935): 38.

86. *Twelfth Census*, cxcvii.

87. ESAR, Wilkinson County, Mississippi, 1935, 6.

88. ESAR, Negro Agent Report, Sumter County, South Carolina, 1929.

89. Victor Perlo, *The Negro in Southern Agriculture* (New York: International Publishers, 1953), 40; Edwin Ware Hullinger, *Plowing Through: The Story of the Negro in Agriculture* (New York: William Morrow, 1940), 22–52, 53–59.

90. ESAR, Annual Narrative Summary for Negro Work, Mississippi, 1935, 1, 47.

91. ESAR, Noxubee County, Mississippi, 1935, 5.

92. J. C. Holton, "Hot Tips for Cold Months," *Southern Agriculturist* 67 (December 1937): 5.

93. ESAR, Oconee County, South Carolina, 1940, 54.

94. J. R. Allgyer, "Artificial Insemination Programs for the Improvement of Livestock," *Proceedings of the Fortieth Annual Convention of the Association of Southern Agricultural Workers, 1939* (Association of Southern Agricultural Workers, n.d.), 94.

95. ESAR, Orangeburg County, South Carolina, 1940, 74.

96. ESAR, Pickens County, South Carolina, 1940, 74.

97. ESAR, E. H. White, Director, Annual Report of the Mississippi Extension Service, 1940, 31–32.

98. ESAR, Walthall County, Mississippi, 1939, 8–9.

99. ESAR, Holmes County, Mississippi, 1940, 27.

100. ESAR, Jefferson County, Mississippi, 1940, 17; ESAR, Lee County, Mississippi, 1940, 8; and ESAR, Newton County, Mississippi, 1940, 30.

101. ESAR, Wilkinson County, Mississippi, 1940, 8.

102. ESAR, State Report for Animal Husbandry, Mississippi, 1940, 20; ESAR, Tippah County, Mississippi, 1940, 18.

103. Nelson, "Welfare Capitalism" 226–32.

104. Horse and Mule Association of America, *Annual Report, 1943*, 3–4.

105. Dacy, "Missouri Mules Make Money" 335.

106. See, for example, ESAR, Forrest County, Mississippi, 1940, 20; ESAR, Greene County, Mississippi, 1940, 6.

107. See, for example, ESAR, Yazoo County, Mississippi, 1940, 5.

Chapter 4

This chapter was originally published in a slightly different form as George B. Ellenberg, "Debating Farm Power: Draft Animals, Tractors, and the United States Department of Agriculture," *Agricultural History* 74 (Summer 2000): 545–68, © 2000 by *Agricultural History*.

1. Tait Butler, "How Should Idle Work Stock?" 30.

2. See Jack Temple Kirby, "The Transformation of Southern Plantations, c. 1920–1960," *Agricultural History* 57 (July 1983): 257–76; Cobb, *Most Southern Place on Earth* 253–54; and Pete Daniel, "Going among Strangers: Southern Reactions to World War II," *Journal of American History* 77 (December 1990): 887–88.

3. Susan Dorothy Jones, "Animal Value, Veterinary Medicine, and the Domestic Animal Economy in the United States, 1890–1930" (PhD diss., University of Pennsylvania, 1997), 186.

4. Derry, *Horses in Society* 233.

5. Stoll, *Larding the Lean Earth* 212.

6. Joe R. Motheral, "The Family Farm and the Three Traditions," *Journal of Farm Economics* 33 (November 1951): 514–29.

7. See Donald H. Grubbs, *Cry from the Cotton: The Southern Tenant Farmers' Union and the New Deal* (Chapel Hill: University of North Carolina Press, 1971); Pete Daniel, *Lost Revolutions: The South in the 1950s* (Chapel Hill: University of North Carolina Press for Smithsonian National Museum of American History, 2000).

8. Deborah Fitzgerald, "Beyond Tractors: The History of Technology in American Agriculture," *Technology and Culture* 32 (January 1991): 116.

9. Deborah Fitzgerald, *Every Farm a Factory: The Industrial Ideal in American Agriculture* (New Haven: Yale University Press, 2003), 110.

10. Robert E. Ankli, "Horses vs. Tractors on the Corn Belt," *Agricultural History* 54 (January 1980): 134–48; Alan L. Olmstead and Paul W. Rhode, "The Agricultural Mechanization Controversy of the Interwar Years," *Agricultural History* 68 (Summer 1994): 35–53.

11. A. G. Smith, *A Farm-Management Study in Anderson County, South Carolina*, USDA Bulletin 651 (Washington, D.C., May 8, 1918), 17.

12. Ibid., 9–10.

13. National Association of Farm Equipment Manufacturers, *Better Living for Farmers* (Chicago: National Association of Farm Equipment Manufacturers, 1925), 5.

14. A. G. Smith, *Farm-Management Study in Anderson County, South Carolina* 10.

15. Ibid., 2–3.

16. Ibid., 16–17. See especially table 7, "Relation of crop area per work animal to farm efficiency."

17. Ibid., 17.

18. Ibid., 21.

19. Ibid., 21–22.

20. HAA, *Proceedings of the First Annual Meeting, 1920* (Chicago: HAA, 1921), 12–13, 35; Olmstead and Rhode, "Agricultural Mechanization Controversy," 39.

21. P. S. Rose to Willet M. Hays, May 13, 1910, Box 18, ROSA-GC.

22. *Progressive Farmer* 35 (January 10, 1920): 12.

23. The committee members were G. A. Bell, Animal Husbandry Division, USDA; S. H. Bell, Bell Bros., importers and exporters of horses; L. W. Chase, University of Nebraska; G. I. Christie, Superintendent of Agricultural Extension, Purdue University; J. B. Davidson, Agricultural Engineering Department, Iowa State University; Wayne Dinsmore, Secretary, American Percheron Society; A. L. Edmonds, Division of Horse Husbandry, University of Illinois; J. I. Falconer, Chief, Department of Rural Economics, Ohio State University; E. J. Gittens, J. I. Case Thrashing Machine Company, Racine, Wisconsin; Howard Green, farmer, Genessee Depot, Wisconsin; W. H. Handschin, Chief, Farm Organization and Management, University of Illinois; E. A. Johnson, International Harvester Corporation; Thomas H. MacDonald, Chief, Bureau of Public Roads, USDA; E. B. McCormick, Chief, Division of Rural Engineering, Bureau of Public Roads, USDA;

H. B. Munger, Chief, Farm Management, Iowa State University; Walter J. Munro, Green, Fulton, Cunningham Company, saddle manufacturers, Detroit, Michigan; Ed E. Parsonage, Deere and Company, Moline, Illinois; F. W. Peck, Cost of Production Section, Office of Farm Management, USDA; G. A. Ranney, International Harvester Company; G. M. Rommel, Chief, Animal Husbandry Division, USDA; H. C. Ramsower, Agricultural Engineering, Ohio State University; E. Rauchenstein, University of Illinois; Lynn S. Robertson, Assistant in Farm Management Demonstration, Purdue University; J. H. Shepperd, Animal Husbandry Department, North Dakota Agricultural College; H. C. Taylor, Chief, Office of Farm Management, USDA; J. V. Taylor, National Hay Association; George Wilbur, farmer, Marysville, Ohio; R. H. Wilcox, Office of Farm Management, USDA; E. A. White, Research and Engineering, Holt Manufacturing Company, Peoria, Illinois; F. A. Wirt, Maryland State College; and J. O. Williams, Animal Husbandry Division, USDA. United States Department of Agriculture, *Proposed Farm Power Studies,* USDA Circular 149 (Washington, D.C., March 1920).

24. Ibid., 7; Farm Power Conference letter, September 16, 1919, Box 663, ROSA-GC.

25. Wayne Dinsmore to Edwin T. Meredith, February 13, 1920, Box 749, ROSA-GC.

26. Susan Dorothy Jones, "Animal Value" 63.

27. HAA, *Proceedings, 1920,* 15.

28. Fred M. Williams to Henry C. Wallace, July 14, 1921, Box 802, ROSA-GC.

29. Wayne Dinsmore to William Jardine, March 21, 1925, Box 1130, ROSA-GC.

30. Fred M. Williams, "Save the Horse Industry and Help Save the Nation" (Chicago: HAA, [1921]), Box 802, ROSA-GC.

31. Edwin T. Meredith to George Massey, April 23, 1920, Box 794, ROSA-GC.

32. Henry C. Wallace to Wayne Dinsmore, June 6, 1921, Box 868, ROSA-GC.

33. Milton S. Eisenhower, "How to Get Agricultural Information," *Southern Agriculturist* 67 (April 1937): 10, 15.

34. H. C. Taylor to F. J. Wright, April 3, 1920, Box 794, ROSA-GC.

35. Henry C. Wallace to Wayne Dinsmore, June 6, 1921, Box 802, ROSA-GC.

36. Wayne Dinsmore to Henry C. Wallace, December 17, 1923, Box 965, ROSA-GC.

37. HAA, "Three to One in Favor of the Mule: Story of the Tractor Demise from the Miller Bros. 101 Ranch Trust, Oklahoma, The Largest Diversified Farm and Ranch in the United States," Leaflet No. 69 (Chicago: HAA, [1920]).

38. Fred M. Williams, "Save the Horse Industry."

39. "Service Rather Than 'Knocking' Propaganda Will Maintain Position of Horse, Truck and Tractor," *Progressive Farmer* 35 (July 17, 1920): 1274.

40. Memorandum from Harlan Smith to [Mr.] Jump, February 9, 1922, Box 898A, ROSA-GC.

41. W. F. Handschin, J. B. Andrews, and E. Rauchenstein, *The Horse and Tractor: An Economic Study of Their Use on Farms in Central Illinois,* University of Illinois Agricultural Experiment Station Bulletin 231 (February 1921): 222–23.

42. HAA, press release, December 7, 1921, Box 898A, ROSA-GC.

43. H. C. Taylor to Finley Mount, January 5, 1922, Box 898A, ROSA-GC.

44. Finley Mount to H. C. Taylor, January 25, 1922, Box 898A, ROSA-GC.

45. F. W. Heiskel to Finley Mount, February 2, 1922, attached to Finley Mount to Henry C. Wallace, February 3, 1922, Box 898A, ROSA-GC.

46. Finley Mount to Henry C. Wallace, February 14, 1922, Box 898A, ROSA-GC.

47. Henry C. Wallace to St. Paul *Pioneer Press*, January 27, 1922, Box 898A, ROSA-GC.

48. Press release attached to Wayne Dinsmore to Henry C. Wallace, March 6, 1922, Box 898A, ROSA-GC.

49. Wayne Dinsmore to Henry C. Wallace, March 6, 1922, Box 898A, ROSA-GC.

50. Henry C. Wallace to Wayne Dinsmore, March 14, 1922, Box 898A, ROSA-GC.

51. Memorandum from Dixon Merrit to [Mr.] Jump, April 26, 1922, Box 898A, ROSA-GC.

52. Ibid.

53. Lynn R. Miller, "Prayers and Precepts," *Small Farmer's Journal* 17 (Spring 1993): 16.

54. Henry C. Wallace to H. C. Taylor, August 15, 1923, Box 984, ROSA-GC.

55. HAA, Reprint of Excerpts from the 1928 Annual Report of the HAA (Chicago: HAA, [1929]), 2, 12–13.

56. Tait Butler, "Use More Horses and Mules," *Progressive Farmer* 39 (April 19, 1924): 6.

57. HAA, Reprint of Excerpts from the 1928 Annual Report, 2, 12–13.

58. HAA, Reprint of Excerpts from the 1929 Annual Report of the HAA (Chicago: HAA, [1930]), 8–9.

59. Wayne Dinsmore to C. W. Warburton, January 18, 1929, Box 1413, ROSA-GC.

60. Wayne Dinsmore to C. W. Warburton, March 21, 1929, Box 1413, ROSA-GC.

61. C. W. Warburton to Wayne Dinsmore, March 28, 1929, Box 1413, ROSA-GC.

62. Wayne Dinsmore to C. W. Warburton, April 2, 1929, Box 1413, ROSA-GC.

63. HAA, Reprint of Excerpts from the 1929 Annual Report of the HAA, 6, Box 1501, ROSA-GC.

64. W. R. Hauser to Wayne Dinsmore, October 31, 1927, Box 1272, ROSA-GC.

65. J. O. Williams, "Horses Used in Big Teams Give Flexible Power and Cut Costs," *Yearbook of Agriculture, 1927* (Washington, D.C., 1928), 384–85.

66. J. T. McAlister, "Annual Report of Extension Work in Agricultural Engineering, December 1, 1929 to November 30, 1930," in ESAR, South Carolina, 1930.

67. Wayne Dinsmore, "The Efficient Use of Animal Power," *Agricultural Engineering* 3 (February 1922): 23.

68. Motheral, "Family Farm" 514–29.

69. Anthony J. Badger, *The New Deal: The Depression Years, 1933–1940* (New York: Noonday Press, 1989), 149.

70. Pete Dyleshorn to Arthur Hyde, November 26, 1931, Box 1656, ROSA-GC.

71. R. M. Evertts to Arthur A. Shaefer, December 13, 1938, Box 2905, ROSA-GC.

72. J. E. Schlottman to Franklin D. Roosevelt, August 18, 1933, Box 1912, ROSA-GC.

73. John Ehresman Jr. to Franklin D. Roosevelt, February 9, 1934, Box 2083, ROSA-GC.

74. Clarence E. Wiles to Henry A. Wallace, January 3, 1934, Box 2083, ROSA-GC.

75. Wayne Dinsmore to Henry A. Wallace, September 16, 1933, Box 1792, ROSA-GC.

76. Charles F. Curtiss to Henry A. Wallace, November 16, 1938, Box 2905, ROSA-GC.

77. Paul S. Taylor to Will Alexander, June 3, 1937, Box 2661, ROSA-GC; Paul S. Taylor to Will Alexander, June 13, 1937, Box 2661, ROSA-GC.

78. C. B. Baldwin to George McGill, July 3, 1933, Box 1792, ROSA-GC.

79. Claude R. Wickard to E. A. Trowbridge, March 27, 1941, Box 294, ROSA-GC.

80. Charles E. Seitz, "Agricultural Engineering Development in Virginia," *Agricultural Engineering* 4 (March 1923): 59.

81. R. W. Trullinger, "Research in Agricultural Engineering," *Agricultural Engineering* 4 (March 1923): 41.

82. Paul Appleby to P. C. Henderson, March 24, 1941, Box 294, ROSA-GC.

83. Mordecai Ezekiel to Morse Salisbury, December 13, 1938, Box 2905, ROSA-GC.

84. Wayne Dinsmore to Arthur M. Hyde, April 30, 1930, Box 1501, ROSA-GC.

85. Ibid.

86. Horse and Mule Association of America, *1940 Annual Report of the Horse and Mule Association of America* (Chicago: HAA, 1940), 11.

87. Edwin T. Meredith to W. W. McCormick, May 3, 1920, Box 757, ROSA-GC.

88. Ezra Taft Benson to Max M. Nicholes, February 16, 1955, Box 2539, ROSA-GC.

89. Arthur M. Hyde to Fred Bohen, June 13, 1930, Box 1501, ROSA-GC.

90. Horse and Mule Association of America, *1940 Annual Report* 6.

91. Clipping attached to A.E. Arthur to Clinton P. Anderson, February 13, 1946, Box 1266, ROSA-GC.

92. Asst. Secretary Hutchinson to H. C. Lodge, January 23, 1952, Box 2076, ROSA-GC.

93. Ezra Taft Benson to Mr. and Mrs. Lewis E. Keylon, January 11, 1956, Box 2766, ROSA-GC.

94. M. H. Berry, "Tractor-Style Tenant Farming," *Southern Agriculturist* 70 (November 1940): 16–17.

95. Ibid.

96. "A Birthday Greeting to Two Million Friends," *Progressive Farmer* 41 (February 13, 1926): 192.

97. "The March of Machinery," *Progressive Farmer* 71 (March 1956): 90.

Chapter 5

1. Derry, *Horses in Society* 40–47.

2. ESAR, Issaqueena County, Mississippi, 1940, 23.

3. Harold Benford, "Power Revolution: The Big Story in Southern Agriculture," *Progressive Farmer* 71 (April 1956): 86.

4. Albert P. Brodell, Paul Stickles, and Paul Wallrabenstein, *Farm Power and Farm Machines* (Washington, D.C., 1953), 7.

5. French, "Raising Colts of Quality" 23.

6. Robert C. Williams, *Fordson, Farmall, and Poppin' Johnny: A History of the Farm Tractor and Its Impact on America* (Urbana: University of Illinois Press, 1987), 131.

7. Dacy, "Missouri Mules Make Money" 336.

8. "Tribute to Real King," *Southern Agriculturist* (May 1938): 3.

9. Arthur Raper, *Machines in the Cotton Fields: Mechanization Comes to the Southern Cotton Farm* (Atlanta: Southern Regional Council, 1946), 89–99. The states and their mule populations were: Mississippi-306,000; North Carolina-269,000; Georgia-257,000; Tennessee-234,000; Alabama-222,000; Texas-191,000; Kentucky-171,000; Arkansas-165,000; South Carolina-159,000; and Louisiana-136,000.

10. William Jackson, "Horses and Mules," *Yearbook of Agriculture, 1943–1947* (Washington, D.C., 1947), 239–41.

11. Greene, "Harnessing Power" 2–3.

12. Dr. Milton P. Jarnagin, "More Colts Needed for Work and Pleasure," *Progressive Farmer* 61 (September 1946): 18.

13. D. W. Walkins, "Ten New Year Hints for Farmers," *Progressive Farmer* 59 (January 1944): 5.

14. "Can Small Farms Make Tractors Pay?" *Progressive Farmer* 61 (October 1946): 83.

15. W. M. Hurst and L. M. Church, *Power and Machinery in Agriculture*, USDA Miscellaneous Publication 157 (Washington, D.C., April 1933), 2–5. See especially tables 1 and 2.

16. Ibid., 2–5; see especially table 3.

17. A. G. Smith, *Farm-Management Study in Anderson County, South Carolina* 10–11.

18. Hurst and Church, *Power and Machinery* 6.

19. Horse and Mule Association of America, Proceedings of the Twentieth Annual Meeting of the Horse and Mule Association of America (Chicago, Horse and Mule Association of America, 1939), 11.

20. Raper, *Machines in the Cotton Fields* 8–9.

21. G. H. Aull, "Changes in the Land of Cotton," *Yearbook of Agriculture, 1958* (Washington, D.C., 1958), 137.

22. Nixon, *Forty Acres* 7.

23. Robert C. Williams, *Fordson, Farmall, and Poppin' Johnny* 131.

24. Harris P. Smith, "What You Can Do: With Mules—With Tractors," *Progressive Farmer* 69 (February 1954): 78, 80.

25. Horse and Mule Association of America, Proceedings of the Twentieth Annual Meeting of the Horse and Mule Association of America, 11.

26. Harris P. Smith, "What You Can Do" 78, 80.

27. Ed Wilborn, "Power Steering for Tractors," *Progressive Farmer* 70 (September 1955): 131.

28. See, for example, Thomas E. Clague, "Should You Use Diesel, LP-Gas, or Gasoline?" *Progressive Farmer* 75 (November 1960): 33.

29. E. L. Langford and B. H. Thibodeaux, *Plantation Organization and Operation in the Yazoo-Mississippi Delta Area*, USDA Technical Bulletin 682 (Washington, D.C.,

May 1939), 58. See table 27 for a detailed view of man labor and power used per acre of cotton.

30. Donald S. Huber and Ralph C. Hughes, *How Johnny Popper Replaced the Horse: A History of John Deere Two-Cylinder Tractors* (Moline, IL: Deere and Company, 1988), 78.

31. J. H. Neal, "Two Mules? Or a Tractor?" *Progressive Farmer* 61 (November 1946): 86.

32. William C. Zelle, "What Form Will the Tractor Ultimately Take?" *Agricultural Engineering* 1 (January 1920): 41.

33. F. A. Wirt, "The General Purpose Farm Tractor," *Agricultural Engineering* 5 (May 1924): 102.

34. "Make the Tractor Do More of the Farm Work," *Progressive Farmer* 35 (March 1940): 724.

35. Robert C. Williams, *Fordson, Farmall, and Poppin' Johnny* 86–87.

36. Dennis S. Nordin and Roy V. Scott, *From Prairie Farmer to Entrepreneur: The Transformation of Midwestern Agriculture* (Bloomington and Indianapolis: Indiana University Press, 2005), 124.

37. C. H. Wendel, *150 Years of International Harvester* (Sarasota, FL: Crestline Publishing, 1981), 259.

38. "Rubber Increases Farm Power," *Southern Agriculturist* 70 (December 1940): 3.

39. Huber and Hughes, *How Johnny Popper Replaced the Horse* 84.

40. Ibid., 59, 93.

41. "A-C's Model B at $495 Aims at a 4,000,000 Farm Market," *Implement and Tractor* 53 (February 5, 1938): 18–19.

42. Robert C. Williams, *Fordson, Farmall, and Poppin' Johnny* 88.

43. Huber and Hughes, *How Johnny Popper Replaced the Horse* 80.

44. Wendel, *150 Years of International Harvester* 346, 349.

45. Hurst and Church, *Power and Machinery* 11–12.

46. Howard Perdue, interview by George B. Ellenberg, June 5, 1990, Special Collections, University Archives, University of Kentucky Libraries.

47. *Implement and Tractor* 52 (July 10, 1937): 15.

48. Booklet on the Ford tractor with the Ferguson system attached to P. C. Henderson to T. J. McAdams, March 5, 1941, Box 294, ROSA-GC.

49. V. J. Dannreuther to Oscar Johnston, April 29, 1933, D&PL-OJ.

50. "Smaller Models for 1938," *Implement and Tractor* 52 (October 2, 1937): 15.

51. M. Taylor Matthews, David R. Jenkins, and Raymond F. Sletto, *Attitudes of Edgefield County Farmers toward Farm Practices and Rural Programs*, South Carolina Experiment Station and USDA Bulletin 339 (February 1942), 6.

52. "Editor's Note," *Progressive Farmer* 61 (October 1946): 84.

53. "Can Small Farms Make Tractors Pay?" 83. See also Neal, "Two Mules? Or a Tractor?" 86–87; and Alexander Nunn, "A 'Two-Horse' Tractor Farm," *Progressive Farmer* 64 (April 1949): 125, 141.

54. Price McLemore, "We Can Save Cotton!" *Progressive Farmer* 63 (February 1948): 62.

55. Arnold P. Yerkes, "Some Aspects of Farm Mechanization," *Agricultural Engineering* (December 1932): 24.

56. National Association of Farm Equipment Manufacturers, *Better Living for Farmers* 11, 14.

57. Allis-Chalmers advertisement, *Southern Planter* (February 1940): 17; Allis-Chalmers advertisement, *Progressive Farmer* 57 (January 1942): 20.

58. See the inside cover of a full-color advertisement for the Farmall Cub, *Progressive Farmer* 62 (August 1947): 2.

59. Qtd. in Harold H. Martin, "Farewell, Jackass!" *Saturday Evening Post* 222 (December 31, 1949): 43.

60. F. M. Johnson, *Veterans Preference for New Farm Machinery and Equipment*, USDA War Records Monograph 4 (Washington, D.C., March 1947), 2–3.

61. See Minneapolis-Moline advertisement, *Progressive Farmer* 70 (February 1955): 171.

62. Harald A. Pedersen, "Mechanization and Its Relation to Farm Labor Changes in the Yazoo-Mississippi Delta," typescript, presented to the Fifteenth Annual Meeting of the Southern Sociological Society, March 27 and 28, 1952, Atlanta, Georgia (Records of the Bureau of Agricultural Economics, Records Relating to Studies Projects, and Surveys, Project Files, 1938–53, Record Group 83, National Archives), 3–4.

63. Ibid., 5–7. See especially table 1, "Operator preferences for tractor or mule operations by level of Mechanization of the Plantation, Bolivar County Survey, 1950."

64. Pedersen, "Mechanization" 8–9.

65. A. P. Brodell and James W. Birkhead, *Age and Size of Principal Farm Machines*, USDA F.M. 41 (Washington, D.C., April 1943), 4.

66. Homer C. McNamara, "Trend in Labor Requirements in Delta Agriculture," August 10, 1940, typescript, D&PL-OJ, 6–7. Lawrence Nelson, who processed these papers initially, noted on this report that McNamara produced this for a Special House Committee investigating tenant displacement, the mechanization of Delta plantations, and mechanization's effect on labor. See page 7 of the typescript for the note and citations of specific letters in the Johnston correspondence file concerning this report.

67. Qtd. in Pedersen, "Mechanization" 9.

68. Pedersen, "Mechanization" 9–11.

69. Nelson, "Welfare Capitalism" 226–32.

70. D&PL-AS (1947), 6.

71. Nelson, "Welfare Capitalism" 226, 244; D&PL-AS (1950), 6; Dunn, *Mr. Oscar* 16.

72. Nelson, "Welfare Capitalism" 230.

73. Oscar Johnston to Progress Manufacturing Company, March 26, 1931, D&PL-OJ.

74. Planters Equipment Company to Oscar Johnston, December 18, 1931, D&PL-OJ.

75. D&PL-AS (1932), 10; Oscar Johnston to Progress Manufacturing Company, March 24, 1931; October 30, 1931; November 11, 1931; and March 14, 1931; D&PL-OJ.

76. Oscar Johnston to J. B. Gully, February 1, 1932, D&PL-OJ.

77. D&PL-AS (1934), 20.

78. Oscar Johnston to Progress Manufacturing Company, March 20, 1931, D&PL-OJ.

79. Oscar Johnston to Minneapolis-Moline Power Implement Company, May 3, 1933, D&PL-OJ.

80. D&PL-AS (1943), 3; D&PL-AS (1944), 3; D&PL-AS (1946), 3; and D&PL-AS (1949), 5.

81. Gilbert C. Fite, *Cotton Fields No More: Southern Agriculture, 1865–1980* (Lexington: University Press of Kentucky, 1984), 167.

82. G. H. Aull, "Changes in the Land of Cotton" 139.

83. D&PL-AS (1951), 4–5; D&PL-AS (1935), 4; D&PL-AS (1938), 1; D&PL-AS (1949), 4–5; and D&PL-AS (1950), 6.

84. Fite, *Cotton Fields No More* 156–57.

85. E. D. White to Charles F. Brannan, February 19, 1948, Box 1583, ROSA-GC.

86. D&PL-AS (1945), 6–7. For a study of Oscar Johnston, see Lawrence J. Nelson, *King Cotton's Advocate: Oscar G. Johnston and the New Deal* (Knoxville: University of Tennessee Press, 1999).

87. Cobb, *Most Southern Place on Earth* 119.

88. E. D. White to Charles F. Brannan, February 19, 1948, Box 1583, ROSA-GC; emphasis added.

89. Martin R. Cooper, *Progress of Farm Mechanization*, USDA Miscellaneous Publication 630 (Washington, D.C., October 1947), 43, 86.

90. Brodell, Stickles, and Wallrabenstein, *Farm Power and Farm Machines* 5.

91. Ibid., 5; Albert P. Brodell and Albert R. Kendall, *Fuel and Motor Oil Consumption and Annual Use of Farm Tractors*, USDA Bureau of Agricultural Economics FM-72 (Washington, D.C., 1950), 2.

92. A. B. Bryan, "Waste Will Raise a Colt for You!" *Progressive Farmer* 58 (February 1943): 10; J. M. Vial, "Six Rules for Horses and Mules," *Progressive Farmer* 60 (March 1945): 22.

93. United States Congress, House Committee on Agriculture, *Farm Machinery Situation* (Washington, D.C., 1948), 205.

94. C. J. McCormick to William Langer, March 9, 1951, Box 1978, ROSA-GC.

95. HAA, Annual Report, 1948, 4.

96. Langford and Thibodeaux, *Plantation Organization* 59.

97. Naum Jasny, "Summary of Talk on the Economics of the Farm Tractor," December 9, 1935, typescript, Records of the Bureau of Agricultural Economics, General Correspondence, 1917–46, Record Group 83, National Archives.

98. Gilbert Fite, "Mechanization of Cotton Production since World War II," *Agricultural History* 54 (January 1980): 191–95.

99. Ernest Stewart, "How Far Mechanized Cotton Farming?" *Progressive Farmer* 63 (October 1948): 82–83.

100. Eugene Butler, "Cotton Mechanizing Fast," *Progressive Farmer* 63 (December 1948): 8.

101. "World's Biggest Cotton Plantation," *Fortune* 15 (March 1937): 160.

102. Qtd. in J. R. Hildebrand, "Machines Come to Mississippi," *National Geographic Magazine* 72 (September 1937): 277.

103. Fite, "Mechanization of Cotton Production" 191.

104. Charles P. Butler and Harold L. Streetman, *Economics of Mechanical Cotton Picking in South Carolina,* South Carolina Agricultural Experiment Station Bulletin 399 (Clemson, South Carolina, January 1952), 34.

105. Oscar Johnston to Orvis Wells, July 22, 1947, Records of the Bureau of Agricultural Economics, Records Relating to Studies, Projects, and Surveys, Record Group 83, National Archives.

106. W. Herbert Brown, *Cotton Farming in the Southern Piedmont, 1930–51,* USDA Bureau of Agricultural Economics Agriculture Information Bulletin 89 (Washington, D.C., June 1952), ii.

107. United States Bureau of the Census, *Graphic Summary of Farm Tenure in the United States, Cooperative Report* (Washington, D.C., 1948), 6.

108. For a discussion of the "push-pull" factors that influenced southern blacks to leave the land, see Donald Holley, *The Second Great Emancipation: The Mechanical Cotton Picker, Black Migration, and How They Shaped the Modern South* (Fayetteville: University of Arkansas Press, 2000), 164–74.

109. For a concise summary of the changes prompted by World War II, see Daniel, "Going among Strangers" 886–911.

110. Howard Perdue interview.

111. Merle C. Prunty and Charles S. Aiken, "The Demise of the Piedmont Cotton Region," *Annals of the Association of American Geographers* 62 (June 1972): 297.

112. Irby Turner and Hugh L. Gary to James F. Byrnes, September 12, 1946, Box 1267, ROSA-GC.

113. Resolution from Columbia County, Central Community, and Winfield-Appling-Leah Community Farm Bureaus to Paul Brown, December 14, 1946, Box 1267, ROSA-GC.

114. Fite, "Mechanization of Cotton Production" 199.

115. N. E. Dodd to Owen Brewster, January 27, 1947, Box 1438, ROSA-GC.

116. As examples of Johnston's interest, see Oscar Johnston to Leroy Percy, March 17, 1927; Burnett S. Miller to Oscar Johnston, September 10, 1936, D&PL-OJ.

117. D&PL-AS (1948), 4.

118. D&PL-AS (1945), 5–6.

119. Oscar Johnston, "Will the Machine Ruin the South?" *Saturday Evening Post* 219 (May 31, 1947): 94–95.

120. Butler and Streetman, "Economics of Mechanical Cotton Picking in South Carolina" 34, 1.

121. Johnston, "Will the Machine Ruin the South?" 98.

122. George B. Nutt, "What Tractors and Equipment Will Suit Your Farm?" *Progressive Farmer* 4 (April 1950): 24.

123. Howard Perdue interview.

124. Martin, "Farewell, Jackass!" 27.

125. Qtd. in Harris P. Smith, "What You Can Do" 78.

126. B. T. Hutchinson to James E. Van Zandt, March 10, 1952, Box 2065, ROSA-GC.

127. William Robert Fuors to James E. Van Zandt, February 26, 1952, Box, 2065, ROSA-GC. See also Thomas E. Berry to Josh Lee, February 2, 1942, Box 642, ROSA-GC.

128. B. T. Hutchinson to James E. Van Zandt, March 10, 1952, Box 2065, ROSA-GC.

129. Wayne D. Rasmussen, "Impact of Technological Change on American Agriculture, 1862–1962," *Journal of Economic History* 22 (December 1962): 578.

130. Henry Regnery, "Richard Weaver: A Southern Agrarian at the University of Chicago," *Modern Age* 32 (Spring 1988): 103.

131. See, for example, Martin, "Farewell, Jackass!" 26–27, 43–44, 46.

132. Peter F. Drucker, "Exit King Cotton," *Harper's Magazine* 192 (May 1946): 476.

Chapter 6

Portions of this chapter were originally published in a slightly different form as George B. Ellenberg, "African Americans, Mules, and the Southern Mindscape, 1850–1950," *Agricultural History* 72 (Spring 1998): 381–98, © 1998 by *Agricultural History*.

1. See Jimmie Lewis Franklin, "Black Southerners, Shared Experience, and Place: A Reflection," *Journal of Southern History* 60 (February 1994): 13.

2. See Harry Crews, *A Childhood: The Biography of a Place* (New York: Harper and Row, 1978) for an example of how the mule played a key role in southern life during the early twentieth century.

3. David M. Potter, "The Enigma of the South," *Yale Review* 51 (October 1961): 142.

4. Ben Robertson, *Red Hills and Cotton: An Upcountry Memory* (Columbia: University of South Carolina Press, 1942), 106.

5. Qtd. in Cobb, *Mississippi Delta* 62.

6. "Tribute to a Real King," *Southern Agriculturist* 68 (May 1938): 3.

7. For example, see ESAR, Tishomingo County, Mississippi, 1935, 39.

8. Qtd. in "Just a Plain Two Horse Farm," in *Such as Us: Southern Voices of the Thirties*, ed. Tom E. Terrill and Jerrold Hirsch (New York: W. W. Norton, 1979), 108.

9. Shields McIlwaine, *The Southern Poor-White from Lubberland to Tobacco Road* (Norman: University of Oklahoma Press, 1939), 233.

10. Betty W. Carter, "Mules in the Delta," *Mules and Mississippi* (Jackson: Mississippi Department of Archives and History, 1980), 37, 39.

11. Typed agreement on letterhead between C. D. Bolton Company and Willis Jones, February 20, 1936, in possession of the author.

12. See G. B. Fiske, "Mules as Prosperity Gauge," *Southern Agriculturist* 67 (April 1937): 22.

13. Qtd. in Billy Skelton, "Tribute to Mule, Vanishing Breed That Once Filled State's Big Need," *Jackson [Mississippi] Clarion-Ledger*, April 4, 1971.

14. Faulkner, *Absalom, Absalom!* 99.

15. William Faulkner, "Barn Burning," in Louis D. Rubin Jr., *The Literary South* (New York: John Wiley and Sons, 1979), 506.

16. David L. Cohn, "Drunk on Cotton," in David L. Cohn, unpublished autobiography, John D. Williams Library, University of Mississippi, Oxford, Mississippi; qtd. in Cobb, *Most Southern Place on Earth* 131.

17. Crews, "Mythic Mule" 21.

18. Qtd. in Theodore Rosengarten, *All God's Dangers: The Life of Nate Shaw* (New York: Alfred A. Knopf, 1975), 113–18, 532, 493. With the publication of the 1975 edition of this book, Rosengarten revealed that Nate Shaw's real name was Ned Cobb.

19. Harris P. Smith, "What You Can Do" 80.

20. Martin, "Farewell, Jackass!" 43.

21. Lee, *With Their Ears Pricked Forward* 8.

22. Susan Dorothy Jones, "Animal Value" 184–85.

23. Howard Perdue interview.

24. Sam Bowen Farm Records, Ledgers for 1921, 1923, and 1930, Special Collections, Clemson University Libraries, Clemson University.

25. Tom E. Terrill and Jerrold Hirsch, *Such as Us* 108.

26. Rosengarten, *All God's Dangers* 461–62.

27. Ibid., 449–50.

28. Crews, *Childhood* 160–62.

29. ESAR, Agricultural Engineering Specialist Statistical Report, South Carolina, 1930. The report gives credit to the *Rice Journal* for publishing this story. For another version, see Thomas D. Clark, *Pills, Petticoats, and Plows: The Southern Country Store* (New York: Bobbs-Merrill, 1944), 242–43.

30. Qtd. in Robert Brent Toplin, "Between Black and White: Attitudes toward Southern Mulattoes, 1830–1861," *Journal of Southern History* 45 (May 1979): 195–97.

31. Joseph Jones, *Agricultural Resources of Georgia* (Augusta, GA: Steam Press of Chronicle and Sentinel, 1861), 5. See also Peter H. Wood, *Black Majority: Negroes in Colonial South Carolina from 1670 through the Stono Rebellion* (New York: W. W. Norton, 1974), 79–85.

32. *Southern Planter* (May 1855): 134; *American Farmer* (October 27, 1820): 242.

33. *Southern Cultivator* (October 1853): 301.

34. Frederick Law Olmsted, *A Journey in the Seaboard States in the Years 1853–1854, with Remarks on Their Economy* (New York: G. P. Putnam's Sons, 1904), 1:51.

35. *Southern Planter* (March 1850): 84.

36. Riley, *Mule* 54.

37. Warder, "Mule Raising" 183–84.

38. Minor Gray interview, May 8, 1974, tape 15–74, transcript, Delta and Pine Land Company Records, Delta and Pine Land Oral History Interviews, Special Collections, Mitchell Memorial Library, Mississippi State University.

39. Cornelius Bostic interview, May 6, 1974, tape 5–74, transcript, Delta and Pine

Land Company Records, Delta and Pine Land Oral History Interviews, Special Collections, Mitchell Memorial Library, Mississippi State University.

40. Hill, "Horse Trader."

41. Edwin A. Hunger, "Let the Cultivator Be Your Handy Man," *Progressive Farmer* 39 (April 1924): 8.

42. Paul K. Conkin, *The Southern Agrarians* (Knoxville: University of Tennessee Press, 1988), 97.

43. Donald Davidson, "Still Rebels, Still Yankees," in *Still Rebels, Still Yankees and Other Essays*, by Donald Davidson (Baton Rouge: Louisiana State University Press, 1957), 239, 248.

44. Clark, "Trade between Kentucky and the Cotton Kingdom" 42.

45. Louis D. Rubin Jr., "The American South: The Continuity of Self-Definition," *The American South: Portrait of a Culture* (Baton Rouge: Louisiana State University Press, 1980).

46. William Faulkner, *The Sound and the Fury* (New York: Random House, 1929), 106–7.

47. Qtd. in Cobb, *Mississippi Delta* 62.

48. West, " 'God's eunuch race' " 3–5.

49. "Julia Henderson Interview," in *TAS*, Georgia Narratives, 3:321.

50. "Caroline Ates Interview," in *TAS*, Georgia Narratives, 3:24.

51. "Robert Kimbrough Interview," in *TAS*, Georgia Narratives, 4:365.

52. Eric Foner, *Reconstruction: America's Unfinished Revolution, 1863–1877* (New York: Harper and Row, 1988), 70–71.

53. Claude F. Oubre, *Forty Acres and a Mule: The Freedmen's Bureau and Black Land Ownership* (Baton Rouge: Louisiana State University Press, 1978), 22–23.

54. "Rev. W. B. Allen Interview," in *TAS*, Georgia Narratives, 3:8–9.

55. "Oxen for the Plow," *Southern Cultivator* (February 1871): 53–54; *Southern Cultivator* (May 1873): 178; *Southern Cultivator* (February 1887): 76; and F. L. Teuton, "Resettlement to Help the South," *Southern Agriculturist* 65 (October 1935): 5.

56. See Martin A. Garrett Jr., "The Mule in Southern Agriculture: A Requiem," *Journal of Economic History* 50 (December 1990): 925–30; Kauffman, "Why Was the Mule Used in Southern Agriculture?"

57. Qtd. in Rosengarten, *All God's Dangers* 529–31.

58. Crews, *Childhood* 32.

59. Herbert N. Richardson, "Black Workers and Their Responses to Work through the Songs They Sang" (EdD diss., Rutgers University, 1987).

60. Qtd. in Richardson, "Black Workers" 139–40.

61. Qtd. in William Reynolds Ferris Jr., "Black Folklore from the Mississippi Delta" (PhD diss., University of Pennsylvania, 1969), 68.

62. Qtd. in Richardson, "Black Workers" 140.

63. Ibid., 141. "Gee" and "haw" are the traditional voice commands for draft animals. "Gee" means that the animal should turn right. "Haw" signals a left turn.

64. Daniel, *Breaking the Land* 9–10; Hullinger, *Plowing Through* 11–12.

65. "King's Last March," *Time* 91 (April 19, 1968): 19.

66. Stephen B. Oates, *Let the Trumpet Sound: The Life of Martin Luther King, Jr.* (New York: Harper and Row, 1982), 497.

67. For a portrait of one black farm family's struggles, see Lisa C. Jones, "40 Acres and a Mule Revisited," *Ebony* 48 (August 1993): 66, 68, 70.

68. Paul M. Gaston, *The New South Creed: A Study in Southern Mythmaking* (New York: Alfred A. Knopf, 1970), 8.

69. Cobb, *Most Southern Place on Earth* 266.

70. Faulkner, "Barn Burning" 502.

71. W. O. Armstrong to Henry A. Wallace, January 27, 1937, Box 2661, ROSA-GC; emphasis added.

72. William J. Fraser, "Live Stock Feeding," *Southern Agriculturist* 65 (February 1935): 43.

73. Crews, *Childhood* 119–20.

74. J. E. Stanford, "The High Cost of Carelessness," *Southern Agriculturist* 68 (September 1938): 5.

75. Earle K. Rambo, "Don't Be Tractor-Killed," *Progressive Farmer* 66 (March 1951): 66.

76. Vernon E. Miller, "Your Tractor Can Kill," *Progressive Farmer* 72 (March 1957): 99.

77. Crews, "Mythic Mule" 21.

78. Anonymous, "A Pious Wish," *Southern Agriculturist* 47 (January 1, 1917): 3.

79. Irwin Russell, "Managing a Mule," *Southern Cultivator* (November 1882): 21. Above this poem was an accompanying illustration with the caption "Dat's What Makes Me 'Spise a Mule."

80. Raper, *Machines in the Cotton Fields* 3, 20.

81. Hurst and Church, *Power and Machinery* 6–7; emphasis added.

82. Fite, *Cotton Fields No More* 143.

83. F. A. Wirt, "The General Purpose Farm Tractor" 102.

84. Andrew Nelson Lytle, "The Hind Tit," in *I'll Take My Stand: The South and the Agrarian Tradition, by Twelve Southerners* (New York: Harper and Brothers, 1930), 211–17.

85. National Association of Farm Equipment Manufacturers, *Better Living for Farmers* 2.

86. Nixon, *Forty Acres* 94–95, 5–6.

87. J. E. Stanford, "The Mechanical Cotton Picker," *Southern Agriculturist* 66 (October 1936): 11.

88. Mrs. Bert H. Walling to Franklin Delano Roosevelt, February 4, 1937, Box 2578, ROSA-GC.

89. Pedersen, "Mechanization" 12–15.

90. D&PL-AS (1947), 8.

91. Pedersen, "Mechanization" 12–15.

92. "Cotton and Cotton Mechanization: What's Going to Happen: An Interview with Harold A. Young," *Progressive Farmer* 64 (October 1949): 19.

93. Virlyn A. Boyd, *Rental Arrangements on Tractor and Non-Tractor Farms in the Southern Piedmont,* Southeastern Regional Land Tenure Committee Publication 6 (South Carolina Agricultural Experiment Station, October 1951), 11, 36.

94. Boyd, *Rental Arrangements* 35–37.

95. L. C. Gray to Secretary of Agriculture, January 14, 1935, Box 1963, ROSA-GC; Frank Tannenbaum to Paul Appleby, December 29, 1934, Box 1963, ROSA-GC.

96. Max N. Tharp to Joe R. Motheral, December 9, 1949, Records of the Bureau of Agricultural Economics, Records Relating to Studies, Projects, and Surveys, Project Files, 1938–53, Box 95, Record Group 83, National Archives.

97. Pedersen, "Mechanization" 15–16.

98. Brown, *Cotton Farming* ii; Harold G. Halcrow, *Agricultural Policy of the United States* (Englewood Cliffs, NJ: Prentice-Hall, 1953), 39.

99. Boyd, *Rental Arrangements* 30–31.

100. Pedersen, "Mechanization" 16–17.

101. Ibid., 16–19.

102. Cobb, *Most Southern Place on Earth* 101.

103. Data sheet from the Staple Cotton Cooperative Association of Greenwood, Mississippi, March 21, 1949, Box 1714, ROSA-GC.

104. Perlo, *Negro in Southern Agriculture* 34, 38.

105. Frank Tannenbaum to Paul Appleby, December 29, 1934, Box 1963, ROSA-GC; emphasis added.

106. Arthur P. Chew, "The Response of Government to Agriculture: An Account of the Origin and Development of Agriculture, on the Occasion of Its 75th Anniversary" (Washington, D.C., November 1937), 90–91, in A. Frank Lever Papers Special Collections, Clemson University.

107. F. L. Teuton, "First Bankhead-Jones Tenant Loan Goes to Alabama Family," *Southern Agriculturist* 68 (March 1938): 15.

108. J. W. Finney, "Resettlement: Helping People Help Themselves," *Southern Agriculturist* 66 (March 1936): 8, 10.

109. See Donald Holley, *Uncle Sam's Farmers: The New Deal Communities in the Lower Mississippi Valley* (Urbana: University of Illinois Press, 1975).

110. Marcus and Segal, *Technology in America* 116.

111. Bidwell and Falconer, *History of Agriculture* 287.

Conclusion

1. Hubbard, *Cotton* 21.

2. "Readers Would Change Title of August Story to That Good Old Mule," *Progressive Farmer* 73 (November 1958): 24.

3. James C. Cobb, *Industrialization and Southern Society, 1877–1984* (Lexington: University Press of Kentucky, 1984), 3.

4. See Kirby, *Rural Worlds Lost.*

5. Wendell Berry, *Gift of Good Land* 111.

6. For various perspectives on the region's future, see, for example, Joe P. Dunn and Howard L. Preston, eds., *The Future South: A Historical Perspective for the Twenty-first Century* (Urbana and Chicago: University of Illinois Press, 1991).

7. See, for example, C. Vann Woodward, *The Burden of Southern History* (Baton Rouge: Louisiana State University Press, 1960).

8. For a discussion of southern rural folk culture, see Wayne Flynt, *Poor but Proud: Alabama's Poor Whites* (Tuscaloosa: University of Alabama Press, 1989).

9. See John Egerton, *The Americanization of Dixie: The Southernization of America* (New York: Harper's Magazine Press, 1974).

10. USDA, *1997 Census of Agriculture,* Volume 1: Part 10, Georgia State-Level Data, Table 16: Tenure and Characteristics of Operator and Type of Organization for All Farms and Farms Operated by Black and Other Races: 1997, 1992, and 1987 (Washington, D.C., 1999), 24.

11. USDA *2002 Census of Agriculture,* Operators by Race, Special Reports, Part 1 (Washington, D.C., April 2005), iv; *Statistical Abstract of the United States: 2004–2005* (Washington, D.C., 2004), 524.

12. Douglas C. Bachtel, "Georgia's Shrinking Farm Population," *Georgia Farmers and Consumers Bulletin* 75 (September 23, 1992): 1, 7; Eric Bates, "Plow up a Storm: The UFO [United Farmers Organization] Story," *Southern Exposure* 16 (Spring 1988): 7.

13. *Statistical Abstract of the United States: 2004–2005* 26.

14. Mark M. Smith, "Listening to the Heard Worlds of Antebellum America," *Journal of the Historical Society* 1 (Spring 2000): 65–99.

15. Daniel, *Breaking the Land* 290–98.

16. Theodore H. Savory, "The Mule," *Scientific American* 223 (December 1970): 104.

17. Daniel, *Breaking the Land* 298.

18. H. R. Tolley and A. P. Brodell, "The Role of Machinery in the Development of the Agriculture of the United States," typescript, Records of the Bureau of Agricultural Economics, Box 32, RG 83, National Archives.

19. "Falsely Accusing the Tractor," *Southern Agriculturist* 70 (November 1940): 3–4.

20. For a full treatment of this theme, see David Danbom, *The Resisted Revolution: Urban America and the Industrialization of Agriculture, 1900–1930* (Ames: Iowa State University Press, 1979).

21. J. A. Evans, "Recollections of Extension History," A. Frank Lever Papers, Speeches Series, Special Collections, Clemson University.

22. Frank L. Owsley, "The Pillars of Agrarianism," *American Review* 4 (March 1935): 535.

23. Ibid., 533–37.

24. Rasmussen, "Impact of Technological Change" 588.

25. Daniel, "Going among Strangers" 886–936.

26. Jonathan M. Wiener, "Class Structure and Economic Development in the American South, 1865–1955," *American Historical Review* 84 (October 1979): 987–91.

27. Allis-Chalmers advertisement, *Southern Agriculturist* 70 (March 1940): 35.

28. Walter John Marx, "Farm Machines and the Good Society," *Commonweal* (December 24, 1948): 173, qtd. in Robert C. Williams, *Fordson, Farmall, and Poppin' Johnny* 134.

29. Davidson, "Still Rebels, Still Yankees" 240–41.

30. Qtd. in Paul S. Taylor to Will Alexander, June 3, 1937, Box 2661, ROSA-GC; enclosed in Will Alexander to Paul Appleby, June 21, 1937, Box 2661, ROSA-GC.

31. James A. Montmarquet, *The Idea of Agrarianism: From Hunter-Gatherer to Agrarian Radical in Western Culture* (Moscow: University of Idaho Press, 1989), 13.

32. Raper, *Machines in the Cotton Fields* 3; emphasis added.

33. Donald Davidson, "Futurism and Archaism in Toynbee and Hardy," in *Still Rebels, Still Yankees and Other Essays* 63.

34. Lee, *With Their Ears Pricked Forward* 132.

35. Potter, "Enigma of the South" 151.

36. Lee, *With Their Ears Pricked Forward* 135.

37. Nordin and Scott, *From Prairie Farmer* xvi.

38. Wendell Berry, *Gift of Good Land* 105.

39. David B. Danbom, "Romantic Agrarianism in Twentieth-Century America," *Agricultural History* 65 (Fall 1991): 9.

40. Victor Davis Hanson, *Fields without Dreams: Defending the Agrarian Idea* (New York: Free Press, 1996), xi.

41. Jeanne S. Hurlbert and Williams B. Bankston, "Cultural Distinctiveness in the Face of Transformation: The 'New' Old South," in *The Rural South since the Civil War*, ed. Douglas R. Hurt (Baton Rouge: Louisiana State University Press, 1998), 187.

Note on Sources

Sources on mules in the American South abound, but one must look carefully since much of it is embedded in materials more broadly related to southern agriculture and culture. Among the most important primary sources are periodicals and government documents. For the nineteenth and twentieth centuries, the agricultural journals provide an excellent starting point for exploring draft-animal use, technology, and the larger questions regarding draft animals, as well as for understanding the context in which southern farmers operated. The *Southern Planter, Southern Cultivator, Southern Agriculturist,* and *Progressive Farmer* are rich mines of material. The advertisements in the journals also highlight areas of change and should not be neglected, especially as farmers became more attuned to the availability of mass-produced goods in the twentieth century. In the case of tractors, advertisements highlight what was available to farmers, the costs of machines in the early years of mechanization, and the arguments for shifting to tractors. Other periodicals such as *De Bow's Review, Agricultural Engineering, Implement and Tractor,* and *Breeder's Gazette* fill out the picture of draft animals and mechanization. The journals, of course, present the view of editors and readers who tended to see machines in a positive light. Also, especially in the case of the South, not everyone read the journals. In the occasional article published in popular journals (such as Pamela Barefoot, "Mules and Memories," *Sandlapper* 12 [1979]: 32–37; Harry Crews, "The Mythic Mule," *Southern Magazine* 1 [1987]: 21–23; and "Machines Come to Mississippi," *National Geographic Magazine* 72 [1937]: 263–318), mules often receive more sympathetic treatment. Still, it is impossible to get a full picture of southern agriculture without using agricultural periodicals.

As the role of government in agricultural research, production, and marketing increased during the later nineteenth century and grew enormously during the twentieth century, there was a parallel increase in the number of reports, studies, and publications on the national and state levels. A researcher could spend a lifetime reading bulletins and reports, and thereby gain an important perspective on what the growing agricultural bureaucracy deemed important to investigate and advocate. On the national level, the *Agricultural Census* is a rich source of statistical information on a variety of topics on regional, state, and county levels. Record Group 16, the Records of the Office of the Secretary of Agriculture, is an important source of policy decisions. In a similar vein, Record Group 83, the Records of the Bureau of Agricultural Economics, illuminates the research side of farm questions. The Bureau of Agricultural Economics was extremely important in studying farm efficiency which, in time, drove investigations that compared draft animals and tractors.

As one moves to the state level, the Extension Service Annual Reports are a crucial source for exploring how national policy decisions percolated to the local level. Each year county agents provided both statistical and narrative reports, often with photographs, to the state extension leadership. Of course, the agents used these reports to justify their positions and funding, but the reports nevertheless provide an important source not only of what programs were implemented on the county level, but how farmers reacted to them; the continuity, success, and failure of initiatives; and the local political dynamics that influenced extension work. One must be careful to consider the fact that extension agents tended to work with the more successful farmers and thus neglected a large segment of the farm population, so the picture is slanted. Where African American county agents were funded, the Negro Extension Reports add another dimension to the extension portrait. In addition, extension reports provide a baseline for comparing programs across the South, which is another means of measuring how certain initiatives, such as local mule production, waxed and waned geographically and chronologically. The Extension Service Annual Reports offer a local, human commentary on a changing rural South.

Government publications must not be neglected in a study of southern agriculture. The United States Department of Agriculture's long-running annual publication, *Yearbook of Agriculture,* summarizes a great deal of information in an easily accessible format. Statistical information, reports from various states and communities, and research articles in each volume provide a wealth of information on national, regional, and state trends. In a more focused way, the thousands of bulletins, reports, and studies generated on the national and state levels present a rich source for historians interested in technological change on American farms. Publications under the USDA rubric (such as W. Herbert Brown, *Cotton Farming in the Southern Piedmont, 1930–51* [Washington, D.C., June 1952]; W. M. Hurst and L. M. Church, *Power and Machinery in Agriculture*

[Washington, D.C., April 1933]; E. L. Langford and B. H. Thibodeaux, *Plantation Organization and Operation in the Yazoo-Mississippi Delta Area* [Washington, D.C., May 1939]; and A. G. Smith, *A Farm-Management Study in Anderson County, South Carolina* [Washington, D.C., May 8, 1918]) exemplify the broad range of publications produced as the nation and the South confronted mechanization and its agricultural and social implications. Mule use per se is seldom the focus of such studies, but information on mules is frequently included in USDA studies of mechanization, farm organization, and cotton production. In a few cases (such as J. O. Williams, *Mule Production* [Washington, D.C., July 1948]), mules are the focus, but more often the researcher must glean information from studies that address other questions. For statistical information on draft animals, the various agricultural censuses provide basic information.

Manuscript collections that reside in archives across the South are an important resource for exploring southern mule use, and they help fill in gaps left by government records. Among the most useful is the large collection of Delta and Pine Land Company records housed at the Mitchell Memorial Library at Mississippi State University. Delta and Pine Land Company was hardly a run-of-the-mill plantation, but for information on mule use and mechanization, it cannot be overlooked because of the richness and breadth of its extant records. The Federal Writers' Project Life History Interviews in the Southern Historical Collection, Wilson Library, at the University of North Carolina are good sources for Depression-era context. The Sam Bowen Farm Records at Clemson University, Special Collections, offers a portrait of a farmer, not a planter, in twentieth-century upstate South Carolina. Collections such as the Rice Family Papers which are available at Mississippi State University's Mitchell Memorial Library offer nuggets pertaining to mules, but often much searching is required to find mention of mules. Broadly speaking, the land grant universities across the South house a wealth of information about their states' respective agricultural worlds.

There are many published primary sources that contain information about mules and, if well indexed, are excellent resources. Various volumes of *The Writings of George Washington from the Original Manuscript Sources, 1745–1799*, ed. John Fitzpatrick (Washington, D.C., 1931–1944) are a good source for exploring Washington's advocacy for mules. *The Papers of Henry Clay* (Lexington: University Press of Kentucky, 1959–84) is similarly a good source for Clay's role in improving mule stock and advocating their use in southern agriculture. Travel accounts such as François André Michaux, *Travels to the Westward of the Allegany Mountains, in the States of the Ohio, Kentucky, and Tennessee, and Re-*

turn to Charlestown, through the Upper Carolinas (London: J. Mawman, 1805) and the well-known Frederick Law Olmsted, *A Journey in the Seaboard States in the Years 1853–1854, with Remarks on Their Economy* (New York: G. P. Putnam's Sons, 1904) allow glimpses into the agriculture and culture of states tied to mule raising, trading, and use. As in the case of other sources, mules may not play a central role in the travel accounts, but the animals are often visible nonetheless. Ray Mathis, Mary Mathis, and Douglas Clare Purcell, eds., *John Horry Dent Farm Journals and Account Books, 1840–1892* (Tuscaloosa: University of Alabama Press, 1977) provides researchers with an Alabama farmer's perspective over several decades. Dent's draft animals play an important role in his account of farm life. The Horse Association of America, also entitled the Horse and Mule Association of America for a time, produced annual reports that are central to understanding how the horse industry responded to the threat presented to draft animals by tractors. From the early 1920s to the late 1940s, the *Annual Report of the Horse Association of America* summarized the state of the horse industry and the organization's fight to keep draft animals on American farms.

Secondary sources on mules and mechanization in the South are not particularly plentiful, but the quality of what does exist is second to none. In the 1980s, three studies were published that provide a much fuller understanding of southern agriculture and its relationship to southern culture. Gilbert Fite, *Cotton Fields No More: Southern Agriculture, 1865–1980* (Lexington: University Press of Kentucky, 1984) portrays mechanization in a positive light. Pete Daniel, *Breaking the Land: The Transformation of Cotton, Tobacco, and Rice Cultures since 1880* (Urbana: University of Illinois Press, 1985) examines the cultures surrounding crops and takes a dimmer view than Fite regarding the impact of mechanization and modernization on rural people and culture. Jack Temple Kirby, *Rural Worlds Lost: The American South, 1920–1960* (Baton Rouge: Louisiana State University Press, 1987) explores the gains and losses incurred as the South moved through the twentieth century. Combined, these works comprise an important watershed in the study of southern agriculture and rural culture. Since the 1980s, works on southern culture have continued. Among the best is Wayne Flynt, *Poor but Proud: Alabama's Poor Whites* (Tuscaloosa: University of Alabama Press, 1989). R. Douglas Hurt, ed., *The Rural South since World War II* (Baton Rouge: Louisiana State University Press, 1998) addresses many aspects of southern change over the last half century. Donald Holley, *The Second Great Emancipation: The Mechanical Cotton Picker, Black Migration, and How They Shaped the Modern South* (Fayetteville: University of Arkansas Press, 2000) is a focused study of a central aspect of mechanization and its impact on rural southerners. James C. Cobb's recent studies, such as *The Most Southern Place on Earth: The Mississippi Delta and the Roots of Regional Identity* (New York: Oxford University Press, 1992) and *Away down South: A History of Southern Identity* (New York: Oxford University Press, 2005), do not focus solely on agriculture but are valuable in how they put the region in perspective.

Some works address mules, mule breeding, and mule use specifically. Anthony Dent, *Donkey: The Story of the Ass from East to West* (London: George G. Harrap, 1972) is a useful overview. The first volume of the *American Jack Stock Stud Book* (Nashville: American Breeders Association of Jacks and Jennets, 1891) contains a summary of jack breeds. L. W. Knight, *The Breeding and Rearing of Jacks, Jennets and Mules* (Nashville: Cumberland Press, 1902) is a useful account of mule raising. Harvey Riley, *The Mule: A Treatise on the Breeding, Training, and Uses to Which He May Be Put* (New York: Dick and Fitzgerald, 1867) is a good starting point for understanding nineteenth-century attitudes toward mules. Clarence Leroy Holmes was among the first scholars to address the role of horses and mules on farms, in "Horses and Mules in American Agriculture, with Special Reference to Their Function in Agricultural Production" (PhD diss., University of Wisconsin, 1921). Robert Byron Lamb, *The Mule in Southern Agriculture* (Berkeley: University of California Press, 1963) is a slim volume, but remains an important treatment of southern mules and sets the stage for other works that attempt to explain why southerners chose mules over horses. Martin A. Garrett Jr., "The Mule in Southern Agriculture: A Requiem" *Journal of Economic History* 50 (1990): 925–30, and Kyle Dean Kauffman, "The Use of Draft Animals in America: Economic Factors in the Choice of an Early Motive Power" (PhD diss., University of Illinois at Urbana-Champaign, 1993) add quantitative expertise to the conversation concerning southern mule use. Most recently, Larry Sawers, "The Mule, the South, and Economic Progress," *Social Science History* 28 (2004): 667–90, shows that the debate is not yet finished about why southerners used mules. While not directly related to the South, John D. Billings, *Hardtack and Coffee: The Unwritten Story of Army Life* (Boston: G. M. Smith, 1887, reprint, Lincoln: University of Nebraska Press, 1993) and Emmett M. Essin, *Shavetails and Bell Sharps: The History of the U.S. Army Mule* (Lincoln: University of Nebraska Press, 1997) illustrate the important role mules have played in American military history.

Southern farmers, black and white, included mules in their oral traditions. William Reynolds Ferris Jr., "Black Folklore from the Mississippi Delta" (PhD diss., University of Pennsylvania, 1969) and Herbert N. Richardson, "Black Workers and Their Responses to Work through the Songs They Sang" (EdD diss., Rutgers University, 1987) treat the significant place of mules in African American culture. Roger Wade West, "'God's eunuch race': The Mule in Southern Literature and Folk Culture" (PhD diss., University of Southern Mississippi, 1996) broadens the scope to include the many literary sources that feature mules. West, to a limited extent, also examines the connections between mules and poor white culture. William Faulkner's many works, of course, often include mules. Reminiscences such as William R. Ferris, *Mule Trader: Ray Lum's Tales of Horses, Mules, and Men* (Jackson: University Press of Mississippi, 1998) and Joshua A. Lee, *With Their Ears Pricked Forward: Tales of Mules I've Known* (Winston-Salem, NC: John F. Blair, 1980) give first-hand accounts of mules and men. Harry Crews, *A Childhood: The Biography of a Place* (New York: Harper and Row, 1978) provides a colorful portrait of growing up in the South. A compelling story of an African American farmer is Theodore Rosengarten, *All God's Dangers: The Life of Nate Shaw* (New York: Alfred A. Knopf, 1974).

There are numerous other articles and monographs that add to our understanding of the agricultural South, some directly related to technology and some that provide background. The *Journal of Southern History* and *Agricultural History* are two of the most important journals that contain relevant articles. The *Journal of Farm Economics* also provides context for understanding how mechanization and farm efficiency standards developed. In particular, Joe Motheral, "The Family Farm and the Three Traditions," *Journal of Farm Economics* 33 (November 1951): 514–29, is an analysis of the various schools of thought swirling within the USDA. Victor Perlo, *The Negro in Southern Agriculture* (New York: International Publishers, 1953) and Edwin Ware Hullinger, *Plowing Through: The Story of the Negro in Agriculture* (New York: William Morrow, 1940) present very different views of progress made by African American farmers in the 1930s and 1940s. Deborah Fitzgerald, *Every Farm a Factory: The Industrial Ideal in American Agriculture* (New Haven: Yale University Press, 2003) examines how efficiency and other industrial ideals came to the fore in the agricultural sector. R. Douglas Hurt, *Agricultural Technology in the Twentieth Century* (Manhattan, KS: Sunflower University Press, 1991) is an important source as well.

Cotton mechanization is central to the story of a changed South, but mechanization of American farms in general is important. Arthur Raper, *Machines in the Cotton Fields: Mechanization Comes to the Southern Cotton Farm* (Atlanta: Southern Regional Council, 1946) and James H. Street, *The New Revolution in the Cotton Economy: Mechanization and Its Consequences* (Chapel Hill: University of North Carolina Press, 1957) detail the enormous impact that machines played in changing the face of the rural South. For a detailed treatment of various aspects of mechanization, see Robert E. Ankli, "Horses vs. Tractors on the Corn Belt," *Agricultural History* 54 (January 1980): 134–48, and Alan L. Olmstead and Paul W. Rhode, "The Agricultural Mechanization Controversy of the Interwar Years," *Agricultural History* 68 (Summer 1994): 35–53. George B. Ellenberg, "Debating Farm Power: Draft Animals, Tractors, and the United States Department of Agriculture," *Agricultural History* 74 (Summer 2000): 545–68, looks at the relationship among tractor proponents, draft-animal advocates, and the USDA. Dennis S. Nordin and Roy V. Scott, *From Prairie Farmer to Entrepreneur: The Transformation of Midwestern Agriculture* (Bloomington and Indianapolis: Indiana University Press, 2005) paints a very positive portrait of modern, scientific agriculture.

In recent years, historians have begun to give draft animals more attention in order to provide a fuller understanding of how humans, animals, and machines influenced one another. John Langdon, *Horses, Oxen, and Technological Innovation: The Use of Draught Animals in English Farming from 1066 to 1500* (Cambridge: Cambridge University Press, 1986) provides historical background. Anne Norton Greene, "Harnessing Power: Industrializing the Horse in Nineteenth Century America" (PhD diss., University of Pennsylvania, 2004) examines horses as technology and explores their relationship to industrialization. Margaret Derry, *Horses in Society: A Story of Animal Breeding and Marketing Culture, 1800–1920* (Buffalo: University of Toronto Press, 2006) approaches horses and horse breeding from a scientific angle and places horses in an international context.

On the subject of agrarian thought and draft animals, few works treat draft horses or mules directly, but clearly draft animals are an important touchstone in many works. Leo Marx, *The Machine in the Garden: Technology and the Pastoral Ideal in America* (New York: Oxford University Press, 1964) is a broad treatment of how America has used the pastoral theme in defining itself. Wendell Berry, *The Unsettling of America: Culture and Agriculture* (San Francisco: Sierra Club Books, 1977) has much to say about farming and has long been a strong critic of agribusiness and all that the term implies. Berry expands on agrarian themes in *The Gift of Good Land: Further Essays Cultural and Agricultural* (San Francisco: North Point Press, 1981). Springing out of a specifically southern context, *I'll Take My Stand: The South and the Agrarian Tradition, by Twelve Southerners* (New York: Harper and Brothers, 1930) presents an alternative to an industrialized southern agriculture which, despite its shortcomings, deserves to be read and pondered. In the chapter "Still Rebels, Still Yankees," in Donald Davidson, *Still Rebels, Still Yankees and Other Essays* (Baton Rouge: Louisiana State University Press, 1957), Davidson presents an important statement of agrarian thought, as does Frank Owsley, "The Pillars of Agrarianism," *American Review* 4 (1935): 529–47. David B. Danbom, "Romantic Agrarianism in Twentieth-Century America," *Agricultural History* 65 (Fall 1991): 1–12, addresses the historical and continuing appeal that agrarianism has for Americans. Victor Davis Hanson, *Fields without Dreams: Defending the Agrarian Idea* (New York: Free Press, 1996) is a firsthand account of the difficulties faced by American farmers. In studying the issues of agricultural modernization, one is confronted by a broad range of responses from the promise of a bright future based on scientific and technological innovations to the long and painful decline of rural culture in the midst of technological change. Balancing these responses requires a careful and sympathetic reading of the various sources on this broad and complex topic.

Index